Backcountry Press Presents

Wildflowers of California's Klamath Mountains

Including the Marble Mountain, Russian & Trinity Alps Wilderness Areas

Ken DeCamp

Editing and scientific review by
Julie Knorr and Julie Kierstead

Dedicated to the memory of Marsha Fickert:

friend, road bike enthusiast, culinary artist,
and behind-the-scenes contributor to this guide,
She is greatly missed.

Oregon violet *(Viola hallii)*

Find this book online at www.backcountrypress.com

Book layout by Kenneth DeCamp and Backcountry Press
All photos unless indicated otherwise by Kenneth DeCamp
Book cover map by Jason Barnes

Front cover images by Ken DeCamp
Back cover map by Jason Barnes

Published by Backcountry Press • Kneeland California
Technical editing and proofing by Michael Kauffmann

ISBN 978-1-941624-13-5
Library of Congress Control Number 2017960640

Kenneth DeCamp
Wildflowers of California's Klamath Mountains
Including the Marble Mountain, Russian &
Trinity Alps Wilderness Areas

MIX
Paper from responsible sources
FSC® C005010
www.fsc.org

BACKCOUNTRY PRESS
Humboldt County, CA

Klamath Mountain Geomorphic Province
Including Subranges

Area Covered by This Guide

Table of Contents

Contents continued:

Introduction

Welcome to the new and improved 2nd edition of our re-titled guide. With this current effort we have been able to add 109 new species, bringing the total illustrated to 629. This sounds like a substantial number but when you consider that there are over 1250 flowering plants cataloged in this area of the Klamath Mountains, we are only about half way there. A recently published checklist of the Vascular Plants of the Trinity Alps by Humboldt State University lists 843 species in the Trinity Alps alone. Of course, a complete photographic manual of all species would require a pack mule to haul around making the effort moot for most hikers and backpackers. As it is, this guide will add a hefty 1 ¼ pounds to your pack unless you own the eBook—in which case the whole thing fits nicely on your smart phone. Being a confirmed bibliophile, I prefer the printed book—there is just something comforting about the printed word and no batteries are required so it won't need a re-charge!

Anyone who has searched for specific wildflowers in the back country knows the difficulties faced. It's no simple task to track something down when you have to pack several miles in only to find that you're either too early or too late for the bloom. Many times I've had to make the hike to a certain location several times over successive days, weeks, or years before I've counted success. Occasionally, however, you hit the bloom just right. I have to say, whether you get your specific photographs or not, the trips are never in vain. How can anyone be disenchanted when spending time in the mountains? And besides, miss one wildflower and chances are excellent you will find something else instead. Between the months of May and November the Klamath Mountains are a wildflower lovers paradise and whether you are a professional or an amateur botanist you will never be disappointed when you visit.

I would like to acknowledge the outstanding advice and help I have received from several individuals with keen eyes and deep knowledge of the wildflowers and the natural world. First and foremost Julie Kierstead & Julie Knorr have been outstanding collaborators and scientific reviewers of this work. Without their botanical identifications this book would not have been possible. Others to thank include Marla Knight, a retired Forest Service botanist, Joaquin Hale, talented botanist to whom experts routinely turn to for IDs and advice on tricky discoveries, and James Lee, a riparian ecologist with the Trinity River Restoration Program who always keeps his eyes out for me when I can't seem to find things on my own! Alice G. Jones who has now passed, was a botanist and forester who took me under her wing when I was just a sprout and encouraged me to "botanize." My wife, kids, and grand-kids who have, in one way or another, become botanical explorers when in the back country with me. To all of these folks I owe a great debt of gratitude.

So—to all of my wildflower friends I say GO FORTH and find those little beauties. But, go forth slowly—do the Patrick McManus mosey (one of my favorite humorists)—and spend as much time looking down as you do looking up. You will be richly rewarded if you do.

Ken DeCamp
Spring 2021

Foreword

The Klamath Mountains are one of the most diverse bioregions in the United States. Robert Whittaker, an ecologist from Cornell University, described the Klamaths as a "great meeting ground" due to its central location on the Pacific Coast, its continuity with other mountain systems, its diverse climate, geology, and topography, its complex disturbance regimes, and its long geological history. While the Native People of the region have known about the region's botanical diversity and thrived in managing this unique botanical landscape, the flora of the region has been overlooked by western science until recently—likely due to the ruggedness and remoteness of the terrain and the lack of nearby universities with graduate botany programs.

Within the Klamath Mountains, the first botanical specimens were collected by the Wilkes Expedition in 1841, and in 1852 Scottish botanist John Jeffries collected and named lodgepole pine in the Siskiyou Mountains and foxtail pine in the Scott Mountains. C. Hart Merriam, Chief of the federal Biological Survey, made a brief sweep across the Klamath Mountains in 1898 after a summer spent around Mount Shasta, and encouraged Alice Eastwood, Curator of Botany at the California Academy of Sciences, to mount a botanical expedition to Canyon Creek, a tributary of the Trinity River. She catalogued over 100 species on her brief trip in 1900, recognizing that most of the herbaceous flora would not be identifiable. A more complete flora, and the development of guidebooks to the local flora, would not be produced for almost another century.

Manuals of the flora of California began to be produced in the 20th century. Willis Jepson's Manual of the *Flowering Plants of California* was published in 1925, but Jepson knew that the flora of northwestern California was underrepresented, and worked with local botanists to expand knowledge of local flora. Philip Munz published his *A California Flora* in 1959, building on the work of Jepson, and a large group of botanists published the *Jepson Manual, Higher Plants of California* in 1993, with a revised edition in 2012. Yet all of these works, with all their detail, were somewhat opaque to the amateur botanist interested in the flora of the Trinity Alps.

In 1974, William Ferlatte published *A Flora of the Trinity Alps of Northern California*, which from its title held the promise of opening up the local flora to a wider audience. But Ferlatte restricted the scope of his flora to elevations above 5000 feet, used primarily his own collections, and added more than 75 species in an appendix that was not integrated into the identification keys of the main text. Furthermore, although close to 100 line drawings were included, they had the stiff, flat look of drawings made from dried herbarium specimens. But soon this would all change.

Alice G. Jones, with degrees in botany and forestry, moved to Weaverville in 1950 with her forest ranger husband, and soon began trekking across the Trinity Alps photographing the local flora. She published *Flowers and Trees of the Trinity Alps* in 1986, which finally made the local flora accessible to the general public. Over 300 species were included, organized by flower color (whitish, yellowish, reddish, bluish) with photographs and descriptions of each species, including type localities. My copy, dog-eared and duct-taped, travelled with me for 15 years, until Ms. Jones re-published the book in 2000 with an appendix that included another 168 species. After she died peacefully at home in 2010, at the age of 98, I thought that further revisions to her book, or completely new photographic guides, would never be published, but how mistaken I was.

In 2017, the first edition of *Wildflowers of the Trinity Alps* was published by Ken DeCamp, with assistance from Julie Kierstead and Julie Knorr. Covering the Trinity

Alps, Russian, and Marble Mountain wildernesses and points in between, it included over 500 species and 700 full color images, and was a little more compact than the Jones book. The photographic quality was superb, and DeCamp even provided a several paragraph guide to wildflower photography. Like the Jones book, the text was organized by flower color, which helped to narrow down the list of species that a particular flower in the field might be. Now, in 2020, before my first edition copy has had a chance to be dog-eared and duct-taped, this second edition has emerged. Containing a few edits from the first edition, and including 109 new species, this edition is the standard for photographic guides, not only for the Trinity Alps, but across California's Klamath Mountains. I hope you will enjoy it as much as I will.

James K. Agee
Professor Emeritus
School of Environmental and Forest Sciences
University of Washington
Seattle, Washington

Notes on this revised book

It was determined, after much discussion, that we needed this guide to be more expansive and thereby more inclusive of the individual areas noted in the 1st edition. Wildflowers have a habit of ignoring artificial boundaries and bloom wherever they darned well please—as long as habitat and other influences are compatible with their particular needs and lifestyles. So to say that this is a guide to the wildflowers of the Trinity Alps is somewhat of a misnomer. It is a guide to a much broader region known as the Klamath Mountains which is a huge chunk of rugged real estate occupying much of Northwestern California. This is a region of varied landscapes, jumbled geologies, confusing weather patterns and home to over 1,250 known flowering plants. Several floristic provinces collide here with the epicenter seemingly hovering over the Scott Mountains.

Whether an amateur or professional, botanizing through these mountains is never boring. With so many out of the way places yet to be explored I have a hunch that many more discoveries will be made. With elevations ranging from 1000' around Shasta Lake to over 9,000' at the summits of Mt. Eddy on the Trinity Divide and Thompson Peak in the Trinity Alps there are nooks and crannies galore.

It was decided that a name change was needed and important to better portray the vast landscape this guide covers. No matter where you wander in these mountains this guide will help you along in your understanding of the rich diversity of flowering plants found here.

Wildflower names

Like a lot of people, I grew up knowing wildflowers, trees, and shrubs by names common to my family but, as I was to learn later on and as my co-authors readily pointed out during the production of this guide, common names are variable. In fact, a particular wildflower can have several common names and these are often shared with other flowers. In order to eliminate this confusion the scientific community waded into the fray and attempted to make some sort of sense out of this dilemma. The upshot is that each and every wildflower has only one scientific name. Just about the time I thought this guide was ready to go to the publisher for the first time the Jepson Manual II was published and California botany was revised once again. Continued research and genetic testing produced significant changes in botanic nomenclature which forced a re-write and re-structuring of our entries which means that, as of this publication date, ours is the only up-to-date field guide for this region. We present the wildflowers by color, structure, habit, and of course, family and species. They are alphabetically arranged according to their scientific name but with a nod to the amateur community the header for each is an arrangement of common names beginning with the most accepted to the sometimes more obscure.

Wildflower color and structure

Several users of the previous edition have expressed some amount of frustration when trying to identify a certain flower by color and/or petal structure. It is important to remember that rarely do species fit neatly into a particular box. We can assume that in broad terms a particular flower will be white but there could be one that is variable with a pink tinge or the one that exhibits more white than blue. Other variability in flowers could even include an individual that is described as having five petals but the one you've found has six! It pays to take a closer look at the flowers in the overall population and find an average. There are several examples in this guide that address the color issue either by displaying photographs of both in the most common section or by displaying separate photographs on pages in the appropriate color sections. The vagaries of Mother Nature will always keep us on our toes—especially those of us who lean toward the amateur or semi-pro side of botany. There are several instances in this guide where flowers are displayed out of order but this is done when it is convenient to keep related species on one page or as near each other as possible making it easier for visual comparison.

Blooming seasons

The mountains covered in this guide occupy a huge chunk of real estate and it would be foolish to think that I have photographed every wildflower. There are just too many physical variables in the natural world to allow for that kind of wishful thinking. Geology, ranges in altitude, exposure, and variable weather conditions make it almost impossible to nail down a definitive date for a particular bloom. Because of climate variability, some shrubs and plants begin as early as December in the lower elevations and continue into late October or even November in the high country. Extreme snow years can put the bloom off by a month or more. Drought years can have the opposite effect—bringing the bloom on early. So, no matter how familiar you think you are with an area, there is always something new and unexpected to find. True familiarity with an area comes with frequent visits and even then it's not always enough. It helps to have a second, or a third, or more sets of eyes along because what is obvious to one person is less so to another. As a help of limited nature I have included the approximate locations and dates of when I found and photographed the wildflowers included in this guide.

Serpentine

The northern half of the Trinity Divide and the easternmost one-third of the Trinity Alps are known as the Red Trinities. These mountains are comprised of highly mineralized rock called serpentine. When first exposed to the elements (say in a roadside cut) it is a pale green to almost black but, because of its high iron content, it rusts when exposed to air and water resulting in soils and rock outcrops that are a distinctive brick red to ocher. Because it is highly mineralized it gives rise to unique plant communities like cobra lily fens and western azalea thickets which are not generally found on soils originating on other geological substrates. A great place to introduce yourself to the serpentine plant community is along the Fen Interpretive Trail at Kangaroo Lake located on the north end of the Trinity Divide off the Gazelle-Callahan Road.

Orchids

Several varieties of orchids are found in this area and they are all beautiful. With a few exceptions they are relatively inconspicuous and hard to find. All orchids require an association with specific soil fungi and should NEVER be disturbed. Digging them up and transplanting them to your garden is a surefire way to kill them. Enjoy them where you find them, take lots of photographs, but whatever you do leave them alone!

Heterotrophic plants

Green plants are considered autotrophs because they photosynthesize—making sugar from water and carbon dioxide. The world of heterotrophic plants is complicated but all have moved away from total energy production from photosynthesis toward obtaining organic carbon either from other living beings or through a symbiotic relationship with a fungus. Heterotrophic plants include parasitic and mycotrophic forms.

Parasitic plants include ground-cones and broomrapes. These species contain no chlorophyll and obtain all nutrients by directly tapping the root system of host plants. Dwarf mistletoes (*Arceuthobium* spp.) are also parasitic. The oak mistletoes (*Phoradendron* spp.) are hemiparasites. These species obtain water and some nutrients from its host tree, but also photosynthesize.

Mycotrophic plants are represented in our area by species in the heath (Ericaceae), orchid (Orchidaceae), and broomrape (Orobanchaceae) families. These plants all obtain nutrients through an intermediary mycorrhizal fungi. Mycorrhizal relationships between fungi and plant are symbiotic, in that the fungi expands the root surface area and increases nutrient and water absorption for the plant while the fungi gets nutrients in return. Mychotrophs exploit this symbiosis by parasitizing the hyphae and the fungi unwittingly feeds the mycotroph. Some species in the heath family have both leafless (heterotrophic) and leaf-bearing (autotrophic) forms.

Insectivorous plants photosynthesize but need an energy boost. They obtain this by trapping insects and digesting their nutrients. Species like *Darlingtonia californica* survive on nutrient poor serpentine soils buy supplementing nutrients in this way.

Because of their unique standing in the plant community, I have set aside a section for regional heterotrophic plants near the end.

Invasive plants

An invasive plant has the ability to thrive and spread aggressively outside its native range. A naturally aggressive plant may be especially problematic when it is introduced to a new habitat. However, since invasive plants also reproduce and spread without human help, they also are naturalized—invasives are a small, but troublesome, sub-category of naturalized plants. Chicory, bachelor buttons, Klamath weed, Scotch broom, and star thistle are problematic invasives. Klamath weed is one that you will encounter along roads and trails and in meadows throughout the high country and seems to be spread in the feed and feces of stock animals.

Similar wildflowers

Several groups of wildflowers look so much alike and share so many common traits that telling them apart can be almost impossible for the amateur and even professional botanists. These problematic species include arnicas, asters, stonecrops, and the small to medium sized blue lupines among others. Stonecrops are of special interest because the Klamath Mountains are the center of distribution and diversity for a group very challenging to tell apart. These beautiful succulents all have basal clusters of thick, fleshy leaves and stems topped with clusters of many five-petaled flowers. The species differ in the shape, size, and attachment of stem and basal leaves, in the color of petals and anthers, and the degree to which the petals spread. Flower color ranges from white to cream yellow and pink. Common species in the book area are cream stonecrop (*Sedum Oregonense*) abundant on the divide around Mt. Ashland and ranging south to the Russians and the Trinity Alps; The Eddys stonecrop (*Sedum kiersteadiae*) prevalent in the Eddys and Scott Mountain then west to Canyon Creek in the Trinity Alps, and Heckner's stonecrop (*Sedum laxum* subsp. *heckneri* - not pictured) in a band from the Trinity/Humboldt County line north through eastern Del Norte and western Siskiyou Counties. Canyon Creek stonecrop (*Sedum paradisum* subsp. *paradisum*) is a rare species known from far eastern Humboldt County, east and north through Trinity County to the mountains north of Shasta Lake in Shasta County. Hybrids occur in the Trinity Alps.

Cultural and historical aspects of wildflowers

Certain plants hold a special fascination for me with regards to their cultural and historical significance. Throughout this guide you will find expanded descriptions of some these plants with regards to these subjects. This has sometimes lead to problems in maintaining the alphabetic order of some species. I've tried to minimize this problem but have found it to be unavoidable in some instances.

Safety in the backcountry

On August 15th, 2021 I had a serious encounter with a black bear on a seldom used trail in the Trinity Alps. A day of high expectations for wildflower and scenic photography was, in an instant, turned upside down. The injuries I suffered were serious and the effects long lasting—although at this writing they are slowly beginning to resolve themselves. The encounter was a mutual surprise for the bear and me and I hold no grudge. Almost all of my life has been spent in the mountains and this was a first for me and in no way alters my plans for getting back out in the future. While I had no idea the bear was so close—sleeping probably because the day was very hot — it would have served me well to pay closer attention to my surroundings. If I had, the day might have turned out differently. As it was I struggled to get back to the trailhead and to a hospital where I underwent surgery and spent several days recovering.

Lesson in all this? Simply stay alert which is good advice for several reasons—rattlesnakes, trailside yellow jacket nests AND bears—all of which you could very well encounter all through these mountains. Above all—don't let fear of an encounter like this keep you from visiting these beautiful places—I am more fearful of riding my bike in traffic on roads and highways than I am of a meeting with a bear.

Parts of a flower:
petals
(corolla)
flower stalk
(pedicel)
filament
anther
stamen
pistil
ovary
style
stigma
sepals
(calyx)
Salmon Mountains wake robin

Bear grass (Xerophyllum tenax)

POKEWEED KNOTWEED
Aconogonon phytolaccifolium 3 or 6 petals
Buckwheat family

This is a tall plant that forms dense thickets in damp and partially shaded sites at higher elevations. Almost all of the plants I photographed were 4′ to 6′ in height. Because it is gregarious, groups appear shrubby. Each plant has a single stem which supports numerous lance-shaped to somewhat sharply ovate leaves emerging from petioles along the stalk. The stalk makes a zig-zag, shifting angles slightly between the petioles. Creamy white flowers bloom in dense clusters at the terminus of each stalk and resemble other buckwheats. I found these in bloom at Upper Wright Lake in the Marbles in late July.

SLIMLEAF ONION, NARROW LEAVED ONION, PAPER ONION
Allium amplectens 3 or 6 petals
Onion family

I've found this onion in bloom in one site only and that was around the edges of a serpentine wetland alongside the road to the Stoney Ridge Trailhead in the Trinity Alps. There was a scattered population of them and all were about 12″ tall - each with large clusters of beautiful white to deep pink flowers at the tops of the leafless stems. It's preferred habitats are vernally wet areas in clay or serpentine. It is endemic to California with scattered occurences throughout the state including along the Trinity River, the Upper Trinity River and on Scott Mountain as well as along Union and Canyon Creeks in the Trinity Alps. I photographed these on Stoney Ridge in June.

WESTERN BISTORT, AMERICAN BISTORT 3 or 6 petals
Bistorta bistortoides
Buckwheat family

This is a fairly tall plant found in alpine and subalpine meadows throughout the area covered by this guide. The white to pinkish white flowers grow in dense clusters at the tops of tall stems. The protruding stamens and anthers give the flowers a decidedly hairy look and are the most noticeable characteristics of this species. I photographed this one on Scott Mountain in July. I've also found it around Lower Albert Lake in the Russians and in the Deadfall Lakes Basin on the Trinity Divide. It is native to California and common and widespread throughout the mountainous regions of the state

POKEWEED KNOTWEED
Aconogonon phytolaccifolium
buckwheat family

SLIMLEAF ONION
Allium amplectens
Onion family

WESTERN BISTORT
Bistorta bistortoides
Buckwheat family

QUEEN'S CUPS, BRIDE'S BONNET, BEAD LILY, CLINTONIA
Clintonia uniflora 3 or 6 petals
Lily family

This beautiful little member of the lily family is a gregarious one found in large groups in shaded and very damp sites trailside throughout this area at mid-elevations. The blooms are white and the entire plant rarely gets any taller than 6". There is only one flower per plant, hence *uniflora*. Late in the season the flowers are replaced by shiny purple to deep blue berries from which it derives one of its common names "bead" lily. The name queen's cups was given in 1903 by Julia W. Henshaw in her book *Mountain Wildflowers of Canada.* She says "it certainly is the queen of all the lovely flower-cups found blooming in the mountain valleys of Western Canada, its white petals making a chalice well worthy of the First Lady of the land" (meaning the Queen of England). I photographed these on the trail to Granite Lake in the Trinities but look for them along the trail to Duck Lake in the Russians as well.

CALIFORNIA LADY'S SLIPPER
Cypripedium californicum 3 or 6 petals
Orchid family

This beautiful little orchid prefers serpentinized soils and is fairly common at lower elevations in the eastern portion of the Trinity Alps and the northernmost portions of the Trinity Divide. It likes water and grows in abundance in marshy areas and next to springs and streams. The inflated lower petal forms the slipper, which is white with light pink markings. The upper sepal and the lateral petals are greenish yellow, extending to the sides like wings, while the other two sepals are joined and located behind the slipper. The overarching column is the same color as the sepals and bears the stigma and anthers. Three to 10 flowers grow in the axils of leaf-like bracts on stems that can reach to 2'. The large photograph was taken next to a spring along the Swift Creek Trail. The inset photograph was taken along the Trinity County side of Highway 3, as it winds over Scott Mountain.

MOUNTAIN LADY'S SLIPPER
Cypripedium montanum 3 or 6 petals
Orchid family

In my experience, this orchid is reclusive and not easy to find. It grows in small, scattered groups often hidden in deeply shaded, mixed conifer and hardwood stands. Its stems are erect and can be 24" or more in height with 1 to 3 slippers on the upper stem. The alternate leaves are lance shaped. The "slipper" is white with beautiful light purple streaks along the edges of the opening and inside. The elongated lateral petals are a deep purplish brown, and twisted. The top sepal, which looks like an upright petal, is purplish brown and lightly streaked with greenish-yellow. The two lateral sepals are partly fused and hang behind the slipper. A large, bright yellow column hangs over and extends into the opening of the slipper and has several reddish brown spots arrayed around the edge. I am not one to toss around the word "lovely" very often but this orchid really is lovely. I shot these photographs at the end of May.

QUEEN'S CUPS
Clintonia uniflora
Lily family

CALIFORNIA LADY'S SLIPPER
Cypripedium californicum
Orchid family

MOUNTAIN LADY'S SLIPPER
Cypripedium montanum
Orchid family

ARROW-LEAF BUCKWHEAT, COMPOSITE BUCKWHEAT, HEARTLEAF BUCKWHEAT

Eriogonum compositum var. *compositum* 3 or 6 petals
Buckwheat family

This is an easy buckwheat to spot because of its size—up to 2′ tall in some places. The stems are stout and the basal leaves an elongated heart shape, dark green above and white and wooly beneath. The creamy white flowers bloom in compact heads or umbels that sit atop several elongated branches which in turn arise from the same spot at the top of the main stem. It likes open rocky areas and meadows. I photographed this one above Big Blue in the Russians but have also found it at Sapphire Lake in the Trinity Alps and alongside roads around Trinity & Lewiston Lakes. **Note**: There is a yellow version of this species illustrated on page 168. This is the only variety of *Eriogonum compositum* present in the book area.

NAKED STEM BUCKWHEAT

3 or 6 petals
Eriogonum nudum
Buckwheat family

Here is a buckwheat you don't even have to get out of your car to see. It is a common road and trailside wildflower that blooms at lower elevations throughout the area covered by this guide. It is a tall (to 2′ or more) and somewhat gangly plant that is loosely branched. Each stem sports a cluster of white flowers at the tips. It prefers dry, open and exposed sites. The stems sometimes turn a bright red in the fall. I photographed these along Highwaty 3 near Lewiston in early October. **Note**: There are three varieties of this species in this area.

GREENE'S BUCKWHEAT

3 or 6 petals
Eriogonum strictum var. *greenei*
Buckwheat family

It was the clusters of ball shaped, creamy white flowers that first caught my eye as I was driving the Salmon River Highway just west of Callahan. They seemed to float above low, rounded mats of small, fuzzy, egg shaped and somewhat fleshy leaves in typical buckwheat fashion. This population favored dry, open, rocky sites just above the highway where I found several plants blooming. For the amateur, buckwheats can be maddeningly difficult to identify but this one is a no brainer. It is the most common mid to high elevation small mat-forming buckwheat with cream to pink colored flowers and white-wooly leaves in this area. **Note**: Another small, white wooly-leaved mat forming buckwheat, *E. ovalifolium*, is also present in the Klamath Mountains.

BEAR VALLEY BUCKWHEAT, BEAR BUCKWHEAT

Eriogonum ursinum 3 or 6 petals
Buckwheat family

The creamy, sometimes yellow, flowers of this pretty little buckwheat bloom in a somewhat compact head atop very long stalks. Some of the flowers may be lightly tinged with pink. The leaf stems in the population I photographed were predominantly red and the slightly hairy leaves were green tinged with red. It grows on a wide range of substrates in exposed dry rocky or gravelly sites. I photographed this population below Hirz Mountain Lookout near Shasta Lake in late May. **Note:** *Eriogonum ursinum* var. *erubescens* or blushing wild buckwheat also occurs in this area. The plant structure is the same but the flowers are definitely tinged pink or maroon.

ARROW-LEAF BUCKWHEAT
Eriogonum compositum var. *compositum*
Buckwheat family

NAKED-STEM BUCKWHEAT
Eriogonum nudum
Buckwheat family

GREENE'S BUCKWHEAT
Eriogonum strictum var. *greenei*
Buckwheat family

BEAR VALLEY BUCKWHEAT
Eriogonum ursinum
Buckwheat family

White flowering fawn lilies in this area: Julie Kierstead notes that the fawn lilies are a confusing group. They hybridize, there are intermediates, and sometimes we see evidence of past hybrid events when one of the parents is no longer nearby. Common traits include a preference for shaded sites, often sunny during early blooming time and shaded after the trees and shrubs leaf. The petals are recurved except for *E. shastense* which has out-facing flowers. The elongated leaves are varying shades of green with white or brown mottling.

CALIFORNIA FAWN LILY
Erythronium californicum — 3 or 6 petals
Lily family

California fawn lily is endemic to Northern California. Flowers range from snow-white to cream and are yellow at the throat, sometimes with a band of orange or reddish brown striations. Unlike var. *roderickii*, the anthers of this one are white or cream colored. I photographed these above French Gulch but I've also seen them along the Ditch Grade between Igo and Platina (SW of Redding) and along Slate Creek on the east side of the Trinity Divide.

CREAM FAWN LILY, LEMON FAWN LILY
Erythronium citrinum var. *citrinum* — 3 or 6 petals
Lily family

Cream fawn lily has a broader distribution than Scott Mountain fawn lily and occurs occasionally in this area. The two are quite similar in appearance with the exception of the anthers which, in this one, are white or cream colored. Flowers are white with a lemon yellow throat surrounded by a slight band of deeper yellow or orange striations. They fade to a light pink or purple as they mature. Despite the similarity in the color of flowers and anthers, cream fawn lily's short style (that structure which connects the ovary to the stigma) is 10mm or less and this distinguishes it from California fawn lily, which has a longer style. I took this photograph near Lowden Ranch above the Trinity River in April.

SCOTT MOUNTAIN FAWN LILY, RODERICK'S FAWN LILY
Erythronium citrinum var. *roderickii* — 3 or 6 petals
Lily family

Scott Mountain fawn lily is endemic to the Trinity River Basin from Scott Mountain to the area around Lewiston and Weaverville. What sets var. *roderickii* apart from var. *citrinum* are the anthers, which vary from lilac to brick red. As the flowers age, the petal tips turn a delicate lilac color. A deep reddish-brown band of striations encircle the base of the petals at the throat. I photographed these along Scorpion Creek in March, but they are well represented throughout the indicated area.

SHASTA FAWN LILY
Erythronium shastense — 3 or 6 petals
Lily family

This species was described in 2015 from the limestone surrounding the McCloud and Squaw Creek Arms of Shasta Lake. The snow white flowers have deep orange centers, somewhat striated in appearance, and tend to age a light purple. The stamens are prominent and topped with deep orange anthers. They bloom at the tops of naked stalks up to 8" in height. The matching pairs of basal leaves are mottled, some with dark reddish-green and others with faint silvery-white. They like deep shade on north facing slopes. I photographed these in the limestone below Hirz Mountain Lookout. **Note:** There is abundant limestone throughout parts of the Klamath Mountains and a lot of it is inaccessible. It is reasonable to believe that more as-yet-undiscovered populations of this species are around.

CALIFORNIA FAWN LILY
Erythronium californicum
Lily family

CREAM FAWN LILY. LEMON FAWN LILY
Erythronium citrinum var. *citrinum*
Lily family

SCOTT MOUNTAIN FAWN LILY
Erythronium citrinum var. *roderickii*
Lily family

SHASTA FAWN LILY
Erythronium shastense
Lily family

PURDY'S FRITILLARY
Fritillaria purdyi 3 or 6 petals
Lily family

Finding this little fritillary reaffirmed my long held belief that, when searching for wildflowers, you can never visit the same area too many times. To my surprise, I found it blooming in a large colony where I've walked dozens of times but never in late March. What initially caught my attention was what seemed like a field of small white stars visible through the trees and at the edges of the old road I was driving. The "stars" turned out to be the snow white edges of the fleshy flower petals. The plants are strongly erect, with stems anywhere from 4" to 10" tall. Ovate leaves crowd the bottom of the stem and there is a single, small, lance shaped leaf midway up. A raceme of one or more bell shaped flowers nod at the tops of the flowerng stems. The mostly snow white petals have yellowish-green centers, streaked with deep maroon spots. Sometimes the flowers are tinged with a very light pink. *Fritillaria purdyi* is a rare plant, restricted to northern California, and is a broad endemic of serpentine soils.

RATTLESNAKE-PLANTAIN
Goodyera oblongifolia 3 or 6 petals
Orchid family

Rattlesnake plantain is a common, early summer bloomer. The flowers are small, white, tubular and alternately attached near the tops of tall, leafless stems. Each egg-shaped leaf of the basal rosette has a distinctive netted white stripe down its center. The upper petals form a hood over the lower petals and the opening created looks like a fish with its mouth open. They like dry, semi-shaded places in older mixed forests at elevations up to 7,000 feet. These photographs were taken on Bolt's Hill above Lewiston Lake and on Slate Mountain on the Trinity Divide.

WHITE HASTINGSIA, RUSH LILY
Hastingsia alba 3 or 6 petals
Century plant family

White hastingsia prefers damp open spaces like meadows, usually standing head and shoulders above anything else. On the grassy slopes around Papoose Lake I've seen flowering stalks approaching 5' in height. Its leaves are elongated and basal, anywhere from 1' to 3'. The flowers grow along a long, slender axis, branched or not, that can be up to 18" long. They are a beautiful translucent white with purplish central vein that lends them a silvery look. I photographed this one in Van Matre Meadows but you will find them in mid to high elevation meadows throughout the Trinities. *Hastingsia serpentinicola* is also common in the book area. **Note:** *Hastingsia* was named for Serranus Clinton Hastings, 1814–1893, first Chief Justice of the Supreme Court of California. He helped publish two volumes of the botany of the Pacific coast and because of this contribution would later be recognized with the plant genus *Hastingsia*. **Reason to revoke the name:** In the 1850s, Hastings managed a forced labor camp and organized murderous "Indian hunts" that killed most of the Native American population of Mendocino County, California, as con firmed by a commission established by Hastings Law School.

PURDY'S FRITILLARY
Fritillaria purdyi
Lily family

RATTLESNAKE PLANTAIN
Goodyera oblongifolia
Orchid family

WHITE HASTINGSIA
Hastingsia alba
Century plant family

LONG-TUBED IRIS, SLENDER IRIS
Iris tenuissima subsp. *tenuissima* 3 or 6 petals
Iris family

Early to mid-spring is iris season in these mountains. They bloom throughout the area and are common alongside roads and trails, often in large populations. A lot of people have a "if you've seen one - you've seen 'em all" attitude towards the beautiful wildflower but that's because they haven't taken the time to get up close and personal with them. Members of this group of wild iris are highly variable in petal structure and coloration, but all have a long corolla tube of 1" to 2½". In some, the petals are wide and streaked with deep reddish-purple and yellow while, in others, the petals are long and narrow and sparsely colored with blue to purplish-red veins and no yellow. All of them are tall, growing to 10" in height. The upper photograph was taken along Swift Creek and the lower one along Stewart's Fork. **Note:** I occasionally face the "exactly what color is it?" dilemma when cataloguing flowers for this book. This iris is one of those. While the underlying color is a variation of white the overlying colors are richly varied–sometimes looking more magenta or purple overall. Maybe I should have a section labled "Best Guess!"

SAND LILY
Leucocrinum montanum 3 or 6 petals
Century plant family

According to most of my botanical manuals, sand lily is more at home in the dry, volcanic soils of eastern Siskiyou and Shasta Counties but, while it may not be common in this neck of the woods, it can occur on sandy, rocky flats almost anywhere throughout this area. The flowers are stark white with long, lance shaped petals surrounding a cluster of prominent stamens topped with orange anthers. Leaves are almost linear, dark green and somewhat folded at the mid-rib. The entire plant rarely exceeds 3" or 4" in height. I photographed this bunch in the mid-1990s in the drier areas near the upper end of Morris Meadows but on subsequent trips was unable to locate any. If anyone using this guide sees and/or photographs any in the Morris Meadows area I would be very interested to know.

LONG-TUBED IRIS
SLENDER IRIS
Iris tenuissima subsp. *tenuissima*
Iris family
Shown are two variations

SAND LILY
Leucocrinum montanum
Century plant family

REDWOOD LILY
Lilium rubescens 3 or 6 petals
Lily family

Drive up any of the roads leading to trailheads on the east side of the Trinity Alps from May to early July and you will find beautiful displays of this amazing lily! The flowers, in large clusters at the tops of tall stems, start out a beautiful white with cinnamon spots, turning a deep wine-purple as they age. The petals are glossy, looking like the have been waxed to a high sheen The smell is delicious and intoxicating–reminiscent of that of western azaleas. It can make anyone not in a hurry want to stretch out in a patch of them and take a nap! I photographed this group along the road to the Granite Peak Trailhead but the road to the Swift Creek Trailhead is a sure bet as well! They are endemic to California with populations noted in the northwestern part of the state from the Bay Area north to near the Oregon border.

CASCADE LILY, PURPLE-FLOWERED WASHINGTON LILY
Lilium washingtonianum subsp. *purpurascens* 3 or 6 petals
Lily family

A subspecies of the more typical Washington lily, this one sports fewer cinnamon spots, and the flowers turn a beautiful deep wine purple in the throats and petal tips as the flowers age. It is also lacking the yellow mid-rib of the typical Washington lily. It occurs at mid-elevations in the Trinity Alps and is sometimes associated with huckleberry oak thickets where the long stems and flowers poke out into sunlight. These photographs were taken along the North Fork of Coffee Creek and Swift Creek. Another good place to see them is on the long Canyon Trail before entering the meadows and just above the Bowerman Meadow cutoff in the Trinity Alps. CalFlora indicates limited distributions in Trinity, Siskiyou, Humboldt and Del Norte Counties with isolated occurances in Mendocino and Lake Counties as well. This is a striking wildflower and can brighten any day in the mountains!

WASHINGTON LILY
Lilium washingtonianum subsp. *washingtonianum* 3 or 6 petals
Lily family

Named for Martha Washington, this lily is, indeed, the first lady of the large lilies. The large white flowers are decorated with small cinnamon spots and bloom in clusters at the tops of very tall stems, sometimes poking out through roadside and trailside brush. There is a yellow mid-rib in the lower half of each petal, and the flowers age white or rarely a very pale pink. I photographed this group just below Hirz Mountain Lookout above Lake Shasta in July. This is not a common lily but occurs over a large geographic area. **Note:** the two subspecies of Washington lily can be difficult to tell apart. Purple-flowered Washington lily tends to occur to the northwest of the Mt. Shasta area and into the Klamath Mountains, and Washington lily tends to occur to the southeast and into the Cascades and Sierra Nevada.

REDWOOD LILY
Lilium rubescens
Lily family

PURPLE-FLOWERED WASHINGTON LILY
Lilium washingtonianum subsp. *purpurascens*
Lily family

WASHINGTON LILY
Lilium washingtonianum subsp. *washingtonianum*
Lily family

WESTERN SOLOMON'S SEAL 3 or 6 petals
Maianthemum racemosum
Butcher's-broom family

I've included two different photographs of this common wildflower... a version that I run into frequently in the high country and a low elevation version commonly found in the foothills all the way to the coast.

The leaves and flowering stalk of the high elevation version are much more compact and shorter than the version found at lower elevations–rarely exceeding 8″ to 10″ in height–and the leaves are thicker and almost leathery to the touch. The flowers are larger and bloom in a thick bunch on a short stalk, often upright. I photographed this one above Upper Albert Lake in the Russian Wilderness in September. You will also find them on the slopes of Monument Peak and around Rush Creek lakes in the Trinity Alps.

The low elevation version grows in damp, shaded areas and blooms in early spring. The plant is elegant with long stems and large, alternate, parallel veined leaves diminishing in size the farther from the base they grow. Some plants reach 3′ or more in length and droop slightly towards the ends. The loosely clustered flowers bloom at the ends of the long stalks. In the fall the leaves turn a beautiful gold with deep, satin red berries. The flower clusters are branched in both forms. I photographed these along the Dog Creek Road (the old Delta Toll Road) near Clear Creek Campground on the Trinity Divide.

STARRY FALSE LILY OF THE VALLEY, STARRY SOLOMON'S SEAL
STARRY SOLOMON'S PLUME
Maianthemum stellatum 3 or 6 petals
Butcher's-broom family

This is a wildflower of wet places–streamsides, swampy meadows and springs. The flowers are snow-white with lance shaped petals that usually curve downward slightly from the calyx. The flowering stems zig-azg between the leaves and are unbranched. The bright green leaves range from 6″ to 10″ long and can be either lance shaped or somewhat oval. The more lance shaped leaves typically clasp the stem and are somewhat folded along the mid-rib. The more oval leaves are not folded. The entire plant can be up to 2′ tall. I photographed these in Parker Meadows along Swift Creek in the Trinity Alps. One of the most beautiful displays I have ever seen is along the seldom used and mostly overgrown Bear Creek Trail used to get to Monument Peak in the Trinity Alps. There were hundreds of plants in a lush meadow next to the trail in August.

WESTERN SOLOMON'S SEAL
Maianthemum racemosum
Butcher's-broom family

STARRY FALSE LILY OF THE VALLEY
Maianthemum stellatum
Butcher's-broom family

FLAT SPURRED PIPERIA, ROYAL REIN-ORCHID

Piperia transversa 3 or 6 petals
Orchid family

Until recently my only encounter with this beautiful, but inconspicuous, little orchid was along a popular trail above Keswick Lake in the Sacramento River Canyon. However, while scrambling around in a patch of manzanita and young pineson the Trinity Divide in late May I encountered it again in the Klamath Mountains. The population above Keswick consisted of two plants. The one on the Divide was solitary and though I spent some time looking I only found the one. It was about 8" tall with tiny 3/16" long white flowers fully opening alternately along the stem. There were two deep-green, small and oval basal leaves. I photographed these along Trinity Mountain Road near Jackass Peak in late May. Cal Flora notes an occurrence near Trinity Center probably along Swift Creek.

ALASKA REIN-ORCHID, ALASKA PIPERIA, SLENDER-SPIRE ORCHID

Piperia unalascensis 3 or 6 petals
Orchid family

Though not small by any means, this member of the orchid family is none-the-less inconspicuous. The flowering stem on this particular plant was about 18" tall and slender with tiny 1/8" long greenish-white flowers arrayed alternately along the upper half. There were three plants in this particular population and though I searched a wide area for more this seemed to be it. The large, deep green basal leaves lying flat to the ground were what initially gave the plant away. The flowering stalk and flowers themselves were almost lost in the clutter of background branches and leaves of surrounding shrubs and trees. I photographed these alongside the old and little used Horse Creek trail in the Trinity Alps in August. CalFlora notes an occurrence along Canyon Creek as well.

WHITE-FLOWERED BOG-ORCHID, BOG CANDLE

Platanthera dilatata var. *leucostachys* 3 or 6 petals
Orchid family

This orchid is a common inhabitant of wet, sub-alpine meadows throughout the area. It's a stand-up plant from 7" to almost 40" tall with tapered leaves and showy snow-white to creamy flowers that can be scattered or thickly clustered along the stalk. Even though the flowers are white and abundant, the smaller plants (which are most common in this area) can often be lost in taller grasses and easily ignored by people in a hurry. The tall ones, while not as common, are easily spotted. This is one wildflower that you should get up close and personal with because they are beautiful and delicate. I photographed the one on the left one in Van Matre Meadows in late July and the gigantic version on the right in Bowerman Meadows in July as well, both in the Trinity Alps.

FLAT SPURRED PIPERIA, ROYAL REIN-ORCHID
Piperia transversa
Orchid family

ALASKA REIN-ORCHID, ALASKA PIPERIA
Piperia unalascensis
Orchid family

WHITE-FLOWERED BOG ORCHID, BOG CANDLE
Platanthera dilatata var. *leucostachys*
Orchid family

FAIRY BELLS, HOOKER'S FAIRY BELLS
Prosartes hookeri 3 or 6 petals
Lily family

There is something enchanting about this wildflower and for me it's the way the little clusters of creamy, bell shaped blossoms hang beneath their canopy of leaves - all but hidden from sight. The prominently veined and slightly egg-shaped leaves are arranged alternately along the slightly drooping main stem which can be up to 24" long. They are common along lower elevation trails throughout the area but you can walk right past them and not even see the flowers. They can be found in their preferred habitat of deeply shaded woods. I photographed these along the Stewart's Fork Trail in early May.

WESTERN LADY'S TRESSES, CREAMY LADY'S TRESSES
Spiranthes porrifolia 3 or 6 petals
Orchid family

There is no mistaking this beautiful and rather elegant member of the orchid family. The yellowish-white, elongated tube shaped flowers bloom in a dense, spiral spike that can reach 24" in height. They prefer vernally damp, open, sunny places where they grow in large colonies. I photographed these on Stoney Ridge in the Trinity Alps in late May. It is a wildflower native to California and fairly widespread throughout the state's mountainous regions.

HOODED LADY'S TRESSES
Spiranthes romanzoffiana 3 or 6 petals
Orchid family

This is another small and beautiful member of the orchid family. While some manuals consider their occurrence uncommon, the opposite is quite true in the wet meadows of the Trinity Alps. The flowers are creamy white and arranged in three rows that spiral upward along a tall, erect stem to a height of about 7" or so. The leaves are basal and up to 5" long. Leaves along the stem are nothing more than bracts. I photographed this one in the meadows above Big Caribou Lake in the Trinity Alps in late August but it can begin its bloom in July and sometimes lasts into early September. I have also found them growing in dry, gravelly sites along the coast just inland from the beaches.

TWISTED-STALK
Streptopus amplexifolius var. *americanus* 3 or 6 petals
Lily family

Twisted-stalk somewhat resembles the fairy bells illustrated above and could be confused with them but for the interesting twist to the thin stalks supporting the flowers and fruit. Each flower stalk emerges just below the leaves and supports one to two blooms which, as I've discovered can be highly variable in color–from white to almost purple. Perhaps the most noticeable thing about it isn't the flower however, but the conspicuous bright red to orange berry-like fruit found in late summer. The leaves are prominently veined and alternate along the main stem which can be about 24" long. Twisted stalk is common in deeply shaded sites throughout the high country, indeed–all the way to the coastal strand. This photograph was taken along Big French Creek in the Trinity Alps in late May. I photographed the berries at the outlet stream below Sugarpine Lake in September.

FAIRY BELLS, HOOKER'S FAIRY BELLS
Prosartes hookeri
Lily family

WESTERN LADY'S TRESSES
Spiranthes porrifolia
Orchid family

HOODED LADY'S TRESSES
Spiranthes romanzoffiana
Orchid family

TWISTED-STALK
Streptopus amplexifolius var. *americanus*
Lily family

FOOTHILL DEATH CAMAS, PANICLED ZIGADENE
Toxicoscordion paniculatum 3 or 6 petals
False-hellebore family

This camas tends to grow in dry, mostly open sites at mid-elevations in the Trinities. The flowers bloom in a fairly loose, multi-branched cluster called a panicle along the upper part of the stem. The creamy white petals form a broad cup, are streaked with orange at the base and have somewhat wavy edges. Stamens are prominent and topped with large white anthers. I photographed these on Scott Mountain in late July but have also found them along the road to the Stony Ridge Trailhead.

MEADOW DEATH CAMAS, BLACK SNAKEROOT, STAR LILY
Toxicoscordion venenosum var. *venenosum* 3 or 6 petals
False-hellebore family

Meadow death camas is common in or near wet meadows from the coastal strand to low- and mid-elevations in the Trinity Alps. The plants can reach to 10″ in height with small white flowers blooming in a tightly bunched formation along the upper third of the stem. The stamens are prominent, each topped with a creamy yellow anther. These plants are poisonous to livestock, especially sheep. **As an interesting side note:** Amy Stewart, in her wonderful book *Wicked Plants - The Weed That Killed Lincoln's Mother & Other Botanical Atrocities* writes that members of the Lewis and Clark expedition became violently ill after possibly eating the roots of this plant, mistaking it for a similar, non-poisonous variety known as blue camas. So, if you find it necessary to follow in their footsteps, do so without eating things you can't identify and otherwise know nothing about! I photographed these in an open meadow area alongside the road to the Tangle Blue Trailhead in mid-June.

WESTERN FALSE-ASPHODEL
Triantha occidentalis subsp. *occidentalis* 3 or 6 petals
False asphodel family

The closer you get to this high country wildflower the prettier it gets. Typically it blooms in large populations and brightens up wet meadows throughout these mountains. The cup shaped snow-white flowers, each with six prominent stamens, bloom in tiered clusters at the tops of 1′ to 1½′ stalks which are covered with rust colored, sticky hairs. Leaves are basal, long, slender and almost grass-like. I shot the main photograph in the meadows at the head of Caribou Lake in late August and the small photograph in the meadow at Upper Albert Lake in the Russian Wilderness in September.

FOOTHILL DEATH CAMAS, PANICLED ZIGADENE
Toxicoscordion paniculatum
False-hellebore family

MEADOW DEATH CAMAS
Toxicoscordion venenosum var. *venenosum*
False-hellebore family

WESTERN FALSE-ASPHODEL
Triantha occidentalis subsp. *occidentalis*
False asphodel family

GIANT WHITE WAKEROBIN, SWEET TRILLIUM
Trillium albidum 3 or 6 petals
False-hellebore family

Trilliums are perennial favorites of mine. I have photographed them all over the world and they never seem to lose their magic. They surprise you in deeply shaded and wet areas, popping up with their showy displays when you least expect it. This is the largest of the trilliums in this area, with erect flowers sitting directly atop the leaves. I've seen these over 16" across with flower petals standing up 4" high. As they age they turn a beautiful purple. There is no need to describe these flowers as they are magnificent and cannot be confused with any other. I photographed these near Lewiston but I've also photographed them around Shasta Lake. They are common throughout these mountains and especially abundant on the coast.

OETTINGER'S TRILLIUM, SALMON MOUNTAINS WAKEROBIN, TRILLIUM
Trillium ovatum subsp. *oettingeri* 3 or 6 petals
False-hellebore family

This little trillium likes to hide in streamside thickets and in the shaded edges of wet meadows at mid-elevations throughout the area and though it is hardly inconspicuous, it can be an unexpected surprise. The snow white flowers stand out in the shadows like little stars. The plants themselves are about 7" tall with three deep green leaves near the top of a naked stalk. A single, three-petaled flower nods on an extended pedicel just above the leaves. I photographed this little beauty where Buck Creek enters the North Fork of Coffee Creek in late May. It is endemic to California with limited distribution in the northwestern part of the state–mainly Trinity and Siskiyou counties. I have also found it in the thickets alongside the Bowerman Meadows Trail and in Union Meadows–all in the Trinity Alps.

WHITE TRITELEIA, WHITE BRODIAEA, FOOL'S ONION 3 or 6 petals
Triteleia hyacinthina
Brodiaea family

This is a flower of wet meadows, vernal pools, and seeps and grows from low to mid-elevation throughout the area covered by this book. It is a tall plant, often reaching 24" in height, topped with a dense cluster of white, bowl shaped flowers with prominent green mid-veins. Occasionally, the flowers are tinged with light purple. The leaves are long and narrow with a prominent ridge extending the length of the back. I photographed these on Musser Hill just off Highway 3 west of Weaverville in June. A particularly beautiful display of this triteleia is in a meadow that has reclaimed a remnant of the old Highway 3 just upstream from the Eagle Creek Campground along the Upper Trinity River. Early in the season the blooms appear like a white cloud floating above the old roadbed.

GIANT WHITE WAKEROBIN
SWEET TRILLIUM
Trillium albidum
False-hellebore family

OETTINGER'S TRILLIUM
SALMON MOUNTAINS WAKEROBIN
Trillium ovatum ssp. *oettingeri*
False-hellebore family

WHITE TRITELEIA
Triteleia hyacinthina
Brodiaea family

INSIDE-OUT FLOWER, REDWOOD IVY 3 or 6 petals
Vancouveria planipetala
Barberry family

The flowers on this plant are so tiny and delicate that most people pass right by without ever seeing them, and that's a shame. The plants range to 2′ tall, are loosely branched and hairy. The leaves are compound, with angular heart-shaped leaflets, each about 1 to 1½″ across. Flowers grow in loose panicles, are six petaled, and ¼″ across or smaller. Each petal has two small yellow anthers near the base. It prefers dry, shaded areas at lower elevations. The leaves of the inside-out-flower are beautiful in autumn when they turn bright red and yellow. I photographed the flowers along the Canyon Creek Trail in late May and the leaves at the Stewart's Fork Trailhead in November.

CALIFORNIA CORN LILY, FALSE HELLEBORE
Veratrum californicum var. *californicum* 3 or 6 petals
False hellebore family

Corn lilies grow commonly throughout the Trinity Alps, the Trinity Divide, and the Russians, often by the hundreds in damp to wet meadows and on slopes following the melting snow. They are among the areas earliest plants to emerge and meadows can be covered with thousands of them. They are beautiful plants and the leaves, in particular, are very photogenic. By early summer the plants can sometimes reach heights of 7′ and are topped with large clusters of white flowers. I shot the large photograph in Red Mountain Measows in August, the leaves in Union Meadows in June and the closeup of the flowers in a meadow below Russian lake in the Russian Wilderness in August. **Note:** *Veratrum insolitum* and *V. viride* also occur in the book area.

BEAR-GRASS
Xerophyllum tenax 3 or 6 petals
False hellebore family

Bear-grass deserves special attention in this guide. First, it is culturally significant to western Native Americans. Native People use the tough, fibrous leaves (which turn white when dried and are easily dyed) when weaving beautiful and intricate designs into their baskets. To increase the production of this basket making material, tribes also introduce fire into their bear-grass gardens. Because it grows in large colonies, bear-grass is hard to miss. When you do see it, realize that these stands have been tended by Native People for thousands of years and other plants preferred for basketry are also growing nearby. The snow-white flowers emerge in compact club-shaped racemes at the tops of stems that can reach 6′ in height. The tough leaves are grass-like, forming mounded bunches at ground level. Shorter leaves emerge all along the flowering stalk. I shot these photographs in July in Horse Creek Meadows in the Trinity Alps but it is common throughout these mountains.

INSIDE-OUT FLOWER
Vancouveria planipetala
Barberry family

CALIFORNIA CORN LILY
Veratrum californicum
var. *californicum*
False-hellebore family

BEAR-GRASS
Xerophyllum tenax
False-hellebore family

MILK MAIDS, TOOTHWORT
Cardamine californica 4 petals
Mustard family

When the whiteleaf manzanita begins its mid-winter bloom, milkmaids are often flowering with them. I've found them blooming in the middle of December in the foothills but mid-to-late January is a better bet. It gets its name from the leaf that, in shape, resembles a dog's canine tooth. Flowers are delicate and white to pale pink. The flowers close up as the sun goes down and open again as the sun comes out. The plants are small, often not more than 6" tall but the flowers occur in large clusters making them easy to spot. Look for them at low elevations alongside trails in deep shade. I photographed these on a hillside above Lowden Ranch above the Trinity River in early April. They are common and widespread.

HEARTLEAF BITTERCRESS
Cardamine cordifolia 4 petals
Mustard family

Anyone who spends time in the backcountry knows that a surprise is always just around the next bend in the trail, hidden away in the rocky outcrop next to camp or buried in the rich foliage of a meadow. This is especially true for those of us on the hunt for things botanical. This little Cardamine was one of those surprises that previously had escaped my notice on a late August trip into Upper Wright Lake in the Marbles. In late July, I found large, populations scattered in wet, meadowy areas around the lake. My attention was initially attracted by the consistent, rectangular arrangement of the 4 white ruffled petals and the heart shaped leaves which were numerous at the base of each tall flowering stalk and scattered farther up. Numerous flowers are concentrated at the tops of the stalks. The petals are lightly veined with green and surround a deep green center and elongated pistil with six yellow-gold anthers.

NUTTALL'S TOOTHWORT
Cardamine nuttallii 4 petals
Mustard family

This small, low-growing toothwort loves open and wet mid-elevation meadows, the banks of streams and springs. The flowers occur in loose clusters at the tops of 5" to 8" stems. They are small and mostly white, with just a hint of pink before they fully open. I shot this photograph in Union Meadows in late May near receding snowbanks along a stream pouring down from the Parker Divide in the Trinity Alps. **Note:** Other *Cardamine* species occur in this area as well, among them–*Cardamine angulata* and *Cardamine cordifolia*–but distinguishing between them is an almost impossible task, especially for an amateur.

SPRING DRABA, WHITLOW GRASS
Draba verna 4 petals
Mustard family

Here's a teeny "bellyflower" that grows in huge masses on relatively dry, exposed hillsides and flats early in the spring. Most people would likely walk right over the top of it without ever stooping to take a look which is a big mistake! It may be small but it is beautiful. Branched flower stalks rise from a tiny rosette of lance or spoon shaped leaves. Each stalk sports one single white flower with a reddish calyx and four deeply lobed petals—so deeply lobed that a casual observer might count eight petals. I photographed these in the Bolt's Hill area above Lewiston Lake.

MILKMAIDS, TOOTHWORT
Cardamine californica
Mustard family

HEARTLEAF BITTERCRESS
Cardamine cordifolia
Mustard family

NUTTALL'S TOOTHWORT
Cardamine nuttallii
Mustard family

SPRING DRABA, WHITLOW GRASS
Draba verna
Mustard family

GLAUCOUS WILLOWHERB, SMOOTH WILLOWHERB
Epilobium glaberrimum — 4 petals
Evening primrose family

Another day–another willow herb–and this one is a beauty. It's an early spring bloomer in the foothills around the west side of Shasta Lake, but late spring along Trinity Mountain Road on the east side of Trinity Lake. It prefers well shaded and moist sites. The plants are somewhat leggy with several thin stalks and multiple flowers. The flowers range in color from snow white to light pink with clusters of stamens and a prominent stigma.

WATER CRESS
Nasturtium officinale — 4 petals
Mustard family

This cress is common in wet, mid-elevation meadows throughout the Trinity Alps. Look for it early in the spring where it often grows in association with marsh yellow cress and Nuttall's toothwort. The flowers cluster in dense heads at the tops of short 6" to 7" stalks and can oftgen be found in large, mounded populations alongside streams, ponds, and other marshy areas. I shot the large photograph in Union Meadows in late May and the small photograph along the Coffee Creek Roads. I have also found them in the wet meadows around Big Carmen Lake in the Scott Mountains.

PENNYCRESS
Noccaea fendleri susp. *glauca* — 4 petals
Mustard family

The flowers in this species bloom in rounded clusters at the tops of the flowering stems. The snow-white petals begin in tubular form but as they mature, they flare slightly away from deep yellow anthers and stamens. Leaves are mostly basal and form a rounded cluster from which the flowering stems emerge. Small alternate leaves are found along the stems. Habitat requirements vary but they prefer seasonally wet rocky areas. I found these blooming beneath an overhanging snowbank on a serpentine talus slope on Scott Mountain in late May.

TOWER ROCKCRESS
Turritis glabra — 4 petals
Mustard family

This somewhat unassuming wildflower can be impressive for one particular reason and that is its height. In the neighborhood of Big Flat and the meadows near Packer's Peak Trailhead I found specimens upwards of 28" tall. Of course, those giants were often surrounded by smaller plants ranging downwards to about 12" more or less. Because of the scattered nature of the individuals and the tiny size of the flowers it probably appears to most passersby as just another weed but get up close and personal with it and the beauty of the flowers becomes apparent. They are snow white, alternate on the stems, long and tubular in presentation with a slight flare at the ends of the petals. Stems and sepals are hairy. The dry fruits are long, linear, and hang down along the stems. I photographed these on the dry flats below Big Flat in late June.

GLAUCOUS WILLOWHERB
Epilobium glaberrimum
Evening primrose family

WATER CRESS
Nasturtium officinale
Mustard family

PENNYCRESS
Noccaea fendleri ssp. *glauca*
Mustard family

TOWER ROCKCRESS
Turritis glabra
Mustard family

WHITNEY'S MILK VETCH, BALLOON POD MILK VETCH, WHITNEY'S LOCOWEED

Astragalus whitneyi — 5 irregular petals

Pea family

The flowers of this member of the pea family are probably not what you will remember about it. Instead it will be the large, inflated white and red pods that look suspiciously like little balloons. Some manuals call this wildflower uncommon but that is not the case throughout the Klamath Mountains. I have found it in several locations in the Trinity Alps, the Scott Mountains and the Marbles. The plants are fairly compact, with stiff stems. Leaves are narrow, lance shaped, alternate, pinnately compound and folded at the margins into long cup shapes. The typical pea-shaped flowers, are creamy white and somewhat tinged with pink–the buds are especially so. It likes open, exposed rocky sites so areas near ridge tops are good places to look for it. I photographed these above Gumboot Lake on the Trinity Divide in early June. The inset photograph of the "balloon" seed pod is provided courtesy of Meredith Parks.

OTTLEY'S STIPULATE LOTUS

Hosackia stipularis var. *ottleyi* — 5 irregular petals

Pea family

On a rainy day in May I was driving to the Granite Peak Trailhead when I spotted clusters of bright white and magenta blooms alongside the road. Leaf and flower structure are both typical of the other lotuses but the color makes it a real standout and could make it confusing to the amateur botanist–is it really white with hints of magenta or is it magenta with hints of white? Obviously, we're going with the former. Look for it in bloom throughout the foothills and mid elevation mountains. If you are attentive you can not miss it.

SICKLE-KEELED LUPINE

Lupinus abicaulis — 5 irregular petals

Pea family

This lupine is locally common along the shoulders and cutbanks of Highway 3 from the Stewart's Fork Bridge and then north, all the way to Trinity Center. A good place to safely check them out is in the vicinity of the intersection of Highway 3 and Bowerman Ridge Road between Covington's Mill and Trinity Center. The plants are tall (around 18" or so) and multi-branched with several stalks of flowers growing on each plant. The keel parts of the flowers are creamy white, while the flags are a light brownish tan, making them somewhat hard to see as you are driving. The flowers can also be purple. The leaves are covered with fine silky hairs. I shot these photograph along Highway 3 in late May

RUSTY SAXIFRAGE

Micranthes ferruginea — 5 irregular petals

Saxifrage family

This delicate little saxifrage grows at higher elevations in the Trinity Alps. I photographed this bunch in August in rocks at the edges of melting snow banks near the top of the pass between Alpine and Smith Lakes and have seen it on the ridge between Salmon and Ward Lakes as well (these are the only places I've ever run across it). The masses of loosely branched red stems will probably catch your eye before the flowers do which, because of their color and diminutive size, blend in quite nicely with the surrounding granite and sandy gravel it grows in.

WHITNEY'S MILK VETCH
Astragalus whitneyi
Pea family

OTTLEY'S STIPULATE LOTUS
Hosackia stipularis var. *ottley*
Pea family

SICKLE-KEELED LUPINE
Lupinus albicaulis
Pea family

RUSTY SAXIFRAGE
Micranthes ferruginea
Saxifrage family

LONG-STALKED CLOVER
Trifolium longipes
Pea family

5 irregular petals

This beautiful clover is found throughout the mountains in this area, often in wet or damp meadow areas. But, I have also found it growing in some unlikely places. A few years ago, while climbing the serpentine rock face at the head of Upper Bowerman Meadows below Lake Anna in the Trinity Alps, I came face to face with several of them blooming in the crevices and small rock shelves a few yards below the lake. The stalks truly are long. I've found them in excess of 16" in Parker Meadows along the Swift Creek Trail. The flowering heads are typical of most clovers. These photographs were taken in early September in the meadow below the Doe Lake outlet in the Trinity Alps and in the meadows above Upper Wright Lake in the Marble Mountains.

Doe Lake in the Trinity Alps is a small lake in a beautiful setting–surrounded by rugged granite peaks and ridges and nearby meadows loaded with wildflowers. A short hike up the trail to the top of the ridge overlooking the lake provides beautiful vistas of Mt. Shasta and the Trinity Divide as well as distant views of the southern portion of the Alps. All-in-all it's a great place to spend a weekend or a week.

LONG-STALKED CLOVER
Trifolium longipes
Pea family

Doe Lake in the Trinity Alps

NORTHERN TWO-EYED VIOLET, WEDGE-LEAVED VIOLET **REVISED**

Viola cuneata

5 irregular petals

Violet family

This is another violet that favors serpentine where it grows in damp, sunny or shaded sites. Like most other violets, it is an early bloomer and can be found, often in abundance, in early to mid-May depending on weather conditions and snowpack at low to mid-elevations. Most of the plants I've found have been less than 6″ in height with basal, heart-shaped, leathery leaves, toothed on the margins. The flowers are small, white tinged with faint purple and deep purple spots on all but the lower petal, which has a deep orange-yellow spot and purple veins extending from the throat. The lateral and two top petals have light yellow in the throat and there is a yellow beard on the two lateral petals. I photographed these along the road to the Lake Eleanor Trailhead in late May. It is also found in serpentine habitat in southern Trinity County as well.

OREGON VIOLET, HALL'S VIOLET

Viola hallii

5 irregular petals

Violet family

This is a violet found on wide open, rocky and sunny serpentine slopes and flats. The plants cluster in small, sprawling mounds with many flowering stems 2″ to 9″ in height. The leaves are deeply dissected with lance-shaped leaflets . The lower petal and two flanking petals are creamy white with a yellow beard and deep yellow spots accented with dark purple veins extending from the throat. The two upper petals are smaller than the others and are a deep burgundy-purple. I photographed these on Scott Mountain in late July where I found large colonies decorating roadside slopes.

MACLOSKEY'S VIOLET

Viola macloskeyi

5 irregular petals

Violet family

This is a meadow violet that favors wet or damp expanses where it often blooms in large colonies. It is small, rarely exceeding 2″ or 3″ in height. The tiny, white flowers are hardly larger than a small fingernail and often obscured from view by grasses and other plants. The lower petal has deep purple veins extending from the throat with a few less obvious veins on the two lateral petals. The lateral petals and the two upper petals have yellow green spots extending outward from the throat with a sparse, almost translucent, beard. The basal leaves are almost round and somewhat cupped. I shot the photograph in Bullard's Meadow along Union Creek in late May.

WESTERN HEART'S EASE, TWO-EYED VIOLET

Viola ocellata

5 irregular petals

Violet family

This is a lower elevation species often found blooming in early May. It is characterized by elongated, heart-shaped leaves and flowering stalks up to 8″ in height. The flowers are stark white with deep purple veins and a yellow-orange spot at the throat on the lower petal. Deep purple spots and yellow beards decorate the two flanking petals just above the throat. The backs of the petals are purple. They seem to like moist, shaded areas and I've frequently found them blooming alongside roads. I shot this photograph near Potem Creek above Shasta Lake in mid-May. While I've never seen it in Trinity or Siskiyou Counties there are records of it from the trails around Gray's Falls and along the New River in the Trinity Alps.

NORTHERN TWO-EYED VIOLET
Viola cuneata
Violet family

OREGON VIOLET, HALL'S VIOLET
Viola hallii
Violet family

MACLOSKEY'S VIOLET
Viola macloskeyi
Violet family

WESTERN HEART'S EASE, TWO-EYED VIOLET
Viola ocellata
Violet family

WESTERN BANEBERRY, RED BANEBERRY, CHINABERRY, DOLL'S EYE
Actaea rubra 5 symmetrical petals - ovary superior
Buttercup family

The creamy-white flowers of this plant are absolutely gorgeous but beware of its beauty. All parts of the plant are poisonous and a healthy adult could experience serious problems after eating as few as 6 of the berries. Two berries could potentially be fatal to a small child. The bright red berries are so bitter though that it is doubtful anyone would be foolish enough to try them. Better safe than sorry though, so keep curious little hands away from the enticing fruit. The deep green compound leaves occur in threes, are coarsely toothed with deeply lobed margins, and are somwhat hairy on the undersides. The flowers bloom in panicles with 3 to 5 petal-like sepals and several prominent stamens. It would be difficult to confuse this wildflower with any other in this area. I photographed the flowers along Stewart's Fork in mid-June and the berries along the shoreline of Sugarpine Lake in early September–both in the Trinity Alps.

BRIDE'S FEATHERS, GOAT'S BEARD, SPAGHETTI FLOWER
Aruncus dioicus var. *acuminatus* 5 symmetrical petals - ovary superior
Rose family

This tall and gregarious plant with its gorgeous, feathery tufts of flowers frequents shaded, wet sites at mid-elevations in this area. In some places I've found them 6' or more in height. Leaves are dark green and deeply divided on the upper 2/3 of the plant. Leaf margins are toothed and the leaflets lance shaped. The flowers are snowy white, numerous, minute, and bloom on long stalks giving it the appearance of feathers which gives rise to the common name Bride's feathers. This is one you can find alongside roads, usually between 2500' and 4000'. I photographed this one along Trinity Mountain Road on the Trinity Divide in mid-June. **Note:** This wildflower is another of those world travellers–I've photographed it in Switzerland where it is called Geissbart.

FIELD CHICKWEED, MEADOW CHICKWEED, FIELD MOUSE-EAR CHICKWEED
Cerastium arvense 5 symmetrical petals - ovary superior
Pink family

This pretty little wildflower grows in thick, clustered mats that can actually cover a lot of ground. Individual plants can grow up to 10" in height and are often covered with hundreds of bright white blossoms. They are especially abundant in wet meadows at mid-elevations. The stems are hairy and sticky to the touch. Opposite, lance-shaped leaves emerge along the entire stem. The flowers are borne in loose clusters and have five deeply cleft petals that flare away from deep green centers. Stamens are topped with light, salmon colored anthers. I shot the large photograph above Kangaroo Lake along the Pacific Crest Trail in July and the smaller one in the meadows below Deadfall lakes on the Trinity Divide also in July. **Note:** This wildflower is another world traveler. I've photographed it meadows in Switzerland where it is called *Acker-Hornkraut*

WESTERN BANEBERRY
Actaea rubra
Buttercup family

BRIDE'S FEATHERS
GOAT'S BEARD
Aruncus dioicus var. *acuminatus*
Rose family

FIELD CHICKWEED
MEADOW CHICKWEED
Cerastium arvense
Pink family

HEARTLEAF SPRING BEAUTY

Claytonia cordifolia
Miner's lettuce family

5 symmetrical petals - ovary superior

The tallish flowering stalks sported several snow-white flowers in various stages of opening extending beyond opposite pairs of heart shaped leaves lower down. The slightly cup-shaped petals are notched at the tips, lightly veined, and surround by an olive-green center with prominent stamens that curve away from the center and lay almost flat to the petals. Each stamen is topped with a pale flesh-colored anther. A late July trip into the Wright Lakes in the Marbles proved productive in the "new-to-me" wildflower department. I was fortunate enough to find several species that I was unfamiliar with including this attractive little *Claytonia*. In late July I found it blooming in large colonies along with heartleaf bittercress in the wet, shaded meadowy areas around Upper Wright Lake.

RYDBERG'S SPRING BEAUTY

Claytonia obovata
Miner's lettuce family

5 symmetrical petals - ovary superior

This little *Claytonia* is one of those wildflowers that can cause an amateur botanist heartburn, and all because of color. In fact, there are other Claytonias that fall into the same category including *C. serpenticola* and *C. sibirica*. Are they white or are they pink? They can be either and often are. On Scott Mountain–where I photographed the *C. obovata* shown here–the flowers are split pretty evenly down the color line and often grow side by side. The leaves emerge in opposite pairs and are oval with pronounced parallel veining. Look for it as snow banks receed on mountain slopes at mid to high elevations. They are gregarious and are found in large populations. **Note:** Research botanist Tommy Stoughton only recently separated *C. obovata* and *C. serpenticola* from their old, long held and commonly accepted botanical classification of *C. lanceolata. C. lanceolata* occurs in the book area. *C. lanceolata* and *C. serpenticola* are similar and can be separated by habitat, with *C. lanceolata* favoring granitics and *C. serpenticola* favoring gabbro and serpentine.

HEARTLEAF SPRING BEAUTY
Claytonia cordifolia
Miner's lettuce family

RYDBERG'S SPRING BEAUTY
Claytonia obovata
Miner's lettuce family

MINER'S LETTUCE
Claytonia perfoliata 5 symmetrical petals - ovary superior
Miner's lettuce family

Miner's lettuce is common and grows at low to mid-elevation throughout Northern California. It prefers moist, shaded areas and often occurs in large, lush colonies. A cluster of small white to pinkish flowers sprouts from what appears to be a single, large, round leaf which is, in fact, two opposite leaves that have grown together. Other leaves are more basal. This miner's lettuce has elliptic to diamond or kidney shaped basal leaves, while the similar *C. parviflora* has linear basal leaves. Occasionally the leaves are mottled with light streaks. Miner's lettuce got its name from early miners who probably learned from Native Americans that eating it would help ward off scurvy in a diet dominated by meat and beans. It has a pleasant taste and is high in vitamin C.

FROSTED MINER'S LETTUCE
Claytonia perfoliata forma *glauca* 5 symmetrical petals - ovary superior
Miner's lettuce family

Another version of miner's lettuce found in the area is known locally as frosted miner's lettuce. Alice Jones refers to it in her book as *Claytonia perfoliata* forma *glauca*. It is typically small with leaves that are mottled with a whitish bloom giving them a silvery look. The white to pink flowers and the leaf structure are similar to *C. perfoliata* only much smaller. Look for this version along the road to the Stewart's Fork Trailhead and again along the lower reaches of the trail where it typically blooms in May and early June.

SERPENTINE SPRING BEAUTY
Claytonia serpenticola 5 symmetrical petals - ovary superior
Miner's lettuce family

If you've ever walked in to Union Meadows or up the Stoney Ridge Trail in late spring or early summer you have undoubtedly seen this gorgeous little wildflower blooming by the thousands. They literally carpet the ground in Union Meadows as snow banks recede. The predominant color in the meadows is pink, or white heavily striped with pink. But up along the Parker Divide and the trail to the Battle Creek Divide the flowers are almost snow white, sometimes with a faint pink striation. The anthers can either be deep rose magenta or faintly pink. The leaves of this spring beauty are decidedly lance shaped and mostly olive green. According to author Tommy Stoughton *C. serpenticola* (recently separated from *C. lanceolata*) is restricted to the ultramafic (serpentenized) soils of the Klamath Ranges of northwest California while *C. lanceolata,* or lanceleaf spring beauty, occurs in the Klamath and favors granitics rather than ultramafics.

CANDYFLOWER, INDIAN LETTUCE, SIBERIAN SPRING BEAUTY
Claytonia sibirica 5 symmetrical petals - ovary superior
Miner's lettuce family

This is a tiny wildflower that frequents damp places like the edges of springs, seeps, and streams at mid-elevations in the high country. The flowering stems rise from a basal rosette of fleshy, oval leaves, are unbranched, and topped with loose clusters of white to pale lavender flowers. The petals are deeply veined. The stamens are topped with pink or lavender anthers. I have photographed this species with six petals, as you will see in the inset. I shot the large photograph alongside Little Deep Creek and the smaller photograph along the road to the Canyon Creek Trailhead, both in the Trinity Alps.

MINER'S LETTUCE
Claytonia perfoliata
Miner's lettuce family

FROSTED MINER'S LETTUCE
Claytonia perfoliata forma *glauca*
Miner's lettuce family

SERPENTINE SPRING BEAUTY
Claytonia serpenticola
Miner's lettuce family

CANDYFLOWER, INIDAN LETTUCE
Claytonia sibirica
Miner's lettuce family

LITTLE PRINCE'S PINE
Chimaphila menziesii — 5 symmetrical petals - ovary superior
Heath family

This little wintergreen grows to about 6" in height on stout, reddish stems. Leaves are oval to elliptic and lightly toothed. Most of the flowers I've encountered have been white or white lightly tinged with pink, and are often solitary but sometimes numbering up to three per stem. The petals curve back slightly, exposing a blunt green stigma in the center. Stamens are prominent, surrounding the green ovary. Look for it in shaded woods. It tends to be more solitary than its larger cousin, prince's pine. I photographed this little bunch on the pass between Bear Basin and Seven-up Lake in the Trinity Alps in mid-September.

ELEPHANT EARS, INDIAN-RHUBARB, UMBRELLA PLANT
Darmera peltata — 5 symmetrical petals - ovary superior
Saxifrage family

Along every river and smallest tributary stream you will find this striking member of the Saxifrage family. I believe it is one of our most iconic Northern California plants. Almost everyone knows at least the leaves because of their size which are quite impressive–sometimes over 24" across. Each bright green leaf sits alone atop a large fleshy stem. The tough, persistent, and thick root stalk clings tightly to submerged rocks and alder roots. The first indications of their presence in the spring are the naked flowering stalks that push their way up and out of the water with dense, flat sprays of small white flowers–each of which has a central cluster of pink pistils. It's the pistils that sometimes give the flowering heads a decidedly pink look. At flowering time, the leaves are just beginning to emerge from the water and by mid-summer they are full grown. In the fall the leaves turn a beautiful gold and red. By mid-winter they are brown, withered and drooping around the stalks before collapsing to the ground. They are beautiful and graceful plants that decorate and shade streambanks throughout the area covered by this book.

This plant has a long history with Native Americans and later with European settlers in these mountains. The young leaf stalk can be harvested, peeled, and eaten raw. It can also be cooked like asparagus or added to soups and stews. **Note:** I do not endorse eating any of the plants found in this guide nor do I endorse any medicinal uses. I do endorse leaving these plants for others to see.

STICKY CINQUEFOIL
Drymocallis glandulosa var. *glandulosa* — 5 symmetrical petals ovary superior
Rose family

There are two color variations of this shrub's flowers–white which is shown here and yellow (page 198). Both are commonly found throughout these mountains. The flowers of this one are creamy white with widely spaced petals backed alternately by the sepals. The stems are decidedly maroon to red. The stems and leaves are deep green, and the leaves prominently veined and toothed at the margins. It can be found blooming on open, mostly dry flats and meadows. I have found this variety in the meadows surrounding the Packer's Peak Trailhead and on dry areas surrounding Big Flat–all in the Trinity Alps. I shot these photographs in late June.

LITTLE PRINCE'S PINE
Chimaphila menziesii
Heath family

ELEPHANT EARS
INDIAN-RHUBARB
Darmera peltata
Rose family

STICKY CINQUEFOIL
Drymocallis glandulosa var. *glandulosa*
Rose family

SISKIYOU RUE-ANEMONE
Enemion stipitatum 5 symmetrical petals - ovary superior
Buttercup family

This is an uncommon and elusive little flower. The delicate white to pale blue "flowers" (the "petals" are actually sepals) are about as big as your little fingernail and bloom at the tops of long, thread-like stalks often nodding to the ground. It's not likely that a casual observer will ever see them. The diminutive size, the scattered nature of the plants, and the fact that they prefer moist areas and deep shade make them hard to find even when you know where they occur. There is a single basal leaf divided into three lobes and then divided into three again. I photographed these near a waterfall in late April in the southern-most portion of the Trinity Divide at about 2000′ in May. You can find them around Weaverville and Lewiston as well.

CAPITATE SANDWORT, BALLHEAD SANDWORT
Eremogone congesta 5 symmetrical petals - ovary superior
Pink family

This scruffy sandwort is found throughout the Klamath Mountains at mid-to higher elevations. It prefers exposed dry and sandy soils and does especially well in rocky crevices or adjacent to rocky outcrops. The flowering stems rise out of a small, basal cushion of bright green needle-like leaves. The flowers bloom in upright, cymose heads with flowers sitting individually atop a short stem. The scruffy nature of the flowers comes from the fact that the petals fold away in a haphazard fashion revealing prominent, thin, and tangled masses of stamens. I shot the large photograph at Emerald Lake and the small inset in the benches above Horse Creek, both in the Trinity Alps .

Lower Russian Lake in the Russian Wilderness. A stiff climb from Waterdog Lake down to this pretty little lake is intimidating for most hikers but if you make it the rewards are hard to beat. The scenery is gorgeous and the area is rich with wildflowers. The steep slope immediately above the lake is especially beautiful when the leopard lilies are blooming but care is required whether climbing down to or back up because footing is precarious.

SISKIYOU RUE-ANEMONE
Enemion stipatum
Buttercup family

CAPTIATE SANDWORT
Eremogone congesta
Pink family

Lower Russian Lake

WOOD STRAWBERRY

Fragaria vesca

5 symmetrical petals - ovary superior

Rose family

If you can find the fruits on this plant consider yourself lucky because critters usually find them first and that is too bad because they are tasty! You might confuse this one with the mountain strawberry but there are definite differences. For one, the leaves on the wood strawberry are more deeply dissected, deeply veined, and not hairy on the edges. The terminal tooth of the central leaflet is equal to the teeth on either side of it. Petals of the wood strawberry usually overlap and the sepals are narrower than in mountain strawberry. I photographed the flowers along the Swift Creek Trail in the Trinity Alps.

MOUNTAIN STRAWBERRY

Fragaria virginiana

5 symmetrical petals - ovary superior

Rose family

The leaves of this strawberry are noticeably blue-green, hairless on the upper surface, but hairy along the toothed edges. Typical of all strawberries, the leaves are divided into three leaflets. In this species, the terminal tooth of the central leaflet is much smaller than the two teeth on either side of it (a major difference between this and wood strawberry). Leaf veins are present but not prominent. The white flowers nestle into the leaves with prominent stamens and flattened oval anthers. It grows in meadows, open woods and along stream banks. The small berries are delicious–if you can beat the birds, chipmunks, and mice to them! This photograph was taken along the Union Creek Trail in late May.

WOOD STRAWBERRY
Fragaria vesca
Rose family

WALD ERDBEERE
"Forest strawberry"
Fragaria virginiana
Rose family

This is the Swiss version of our tasty Mountain strawberry and just as hard to come by when ripe.

MOUNTAIN STRAWBERRY
Fragaria virginiana
Rose family

CARROT-LEAVED HORKELIA
Horkelia daucifolia 5 symmetrical petals - ovary superior
Rose family

This absolutely beautiful *Horkelia* blooms in large colonies in open, sunny sites seeming to favor serpentinized soils. The flowering stalks, which can stand upright or lie flat on the ground, are a bright maroon-red and hairy. The 11-21 narrowly oval leaflets are so hairy that they appear silvery. Flowers bloom in compact clusters, but unlike *H. tridentata*, the snow-white petals of *daucifolia* emerge narrowly from the center, flare into an almost spatulate shape, and are slightly cupped with notched tips. The yellowish-green ovary is surrounded by a ring of prominent, soft yellow stamens and anthers. The flowers are small, so to truly appreciate how beautiful they are you need to get down on your hands and knees for a good long look. I photographed these on Scott Mountain in July.

CARROT-LEAFED HORKELIA
Horkelia daucifolia var. *daucifolia* 5 symmetrical petals - ovary superior
Rose family

The flowers are the same as those of *H. daucifolia*, but the plant itself is different. The leaf structure is somewhat similar, being hairy, but with leaflets far more sparse, or spread apart. The big difference is that the flowering stalk is almost naked and strongly upright with the flowers blooming in close bunches at the tops. Small leaves emerge every so often along the stalk. Driving the Gazelle to Callahan Road I spotted an entire hillside in bloom with this new-to-me horkelia. I walked an area covering about two acres alongside the road to verify the overall plant structure and habitat. In this area they favored disturbed sites like cut banks and were growing more thickly near the road and thinning out considerably farther away. I shot these photographs in mid-June.

THREETOOTH HORKELIA
Horkelia tridentata 5 symmetrical petals - ovary superior
Rose family

This gregarious little wildflower is a sprawler. While some of the flowering stalks stand upright, most hug the ground with hairy red stems. The deep green leaves have 5-11 oval leaflets which are hairy and sharply notched at the tips. The flowers are small, about 1/4" to 3/8" across, with five widely spaced, narrowly oval petals that surround and flare away from a central ring of stout stamens sporting golden brown to deep red anthers. They are common in open sunny places but I've found them to be especially abundant on serpentine soils. I photographed these along the lower reaches of the Stoney Ridge Trail in May.

CARROT-LEAVED HORKELIA
Horkelia daucifolia
Rose family

CARROT-LEAFED HORKELIA
Horkelia daucifolia var. *daucifolia*
Rose family

THREETOOTH HORKELIA
Horkelia tridentata
Rose family

PICKERING'S IVESIA, SILKY MOUSETAILS
Ivesia pickeringii
Rose family

5 symmetrical petals - ovary superior

The name "silky mousetails" is quite fitting, for the leaves are long, soft, and flexible to the touch. They grow in compact, erect bunches 8" to 10" long, narrowly tapered leaves formed by overlapping, lobed leaflets. In fact, the leaves are probably the first thing you will notice about this plant. The flowers are inconspicuous and borne in loose clusters at the tops of 10" to 15" thin, reddish stems and are about 3/8" in diameter. The white or pink tinged petals surround 20 rather large, soft yellow stamens, are narrow at the base then curve away becoming almost round with slightly wavy tips. The sepals are deep maroon tinged and everything–leaves, stalks and sepals–is covered with fuzzy white to gray hairs. This is an uncommon wildflower endemic to the Klamath Mountains and serpentine soils in particular. It favors sunny, dry mountain meadows where I've found it blooming alongside skullcaps and carrot-leaved horkelia. I photographed these on Scott Mountain in late June and have also found them below Stonewall Pass in late Jul

PARTRIDGE FOOT
Luetkea pectinata
Rose family

5 symmetrical petals - ovary superior

This beautiful little wildflower is particularly fond of rocky cliff faces and scree slopes where it forms dense, verdant green mats that hug and encircle the rocks. Its fan shaped, deeply dissected leaves are crowded at the base of the plant (resembling the way the feathers surround partidge's feet). The flowering stalks are nearly leafless. White flowers with deep buttery yellow centers emerge alternately all along the 8" to 10" stem but are most crowded at the top. Stamens and anthers are prominent. In the fall, the flowering stalks, seeds still attached, turn a beautiful rusty brown, which is quite beautiful in contrast with the deep green of the leaves and the surrounding rocks. I photographed this one above Horse Creek in the Trinity Alps but they are found throughout the region at high elevations.

LARGE LEAVED SANDWORT, LARGELEAF SANDWORT
Moehringia macrophylla
Pink family

5 symmetrical petals - ovary superior

Even though this wildflower grows in profusion in certain areas you really need to get down in the dirt to appreciate it. The plants are typically only 1" to 1¼" tall. Leaves are 1" to 2" long, opposite and lance shaped. The tiny flowers (3/16" across) grow in groups of two to five in loose, terminal clusters. The snow-white petals are recurved so that they reveal the light green ovary and stigma. The pale green stamens are prominent and topped with light salmon colored anthers. I photographed these growing in dry, rocky soil along the Stewart's Fork Trail near Little Deep Creek in mid-May.

BROOK SAXIFRAGE
Micranthes odontoloma
Saxifrage family

5 symmetrical petals - ovary superior

This is a relatively common saxifrage found alongside streams, the shorelines of lakes, and wet-meadows at higher elevations throughout the area. It is easily recognized by its evenly toothed and rounded leaves that can be up to 3" across. The tiny flowers are loosely clustered near the top of a tall stalk and each petal has two tiny greenish-yellow dots near the base. I photographed these in the meadow below Mirror Lake in the Trinity Alps in late August.

PICKERING'S IVESIA, SILKY MOUSETAILS
Ivesia pickeringii
Rose family

PARTRIDGE FOOT
Luetkea pectinata
Rose family

LARGE LEAVED SANDWORT
Moehringia macrophylla
Pink family

BROOK SAXIFRAGE
Micranthes odontoloma
Saxifrage family

TOAD LILY, WATER MINER'S LETTUCE, SPRING BEAUTY

Montia chamissoi — 5 symmetrical petals ovary superior

Miner's lettuce family

This is another of the myriad "belly flowers" you will find illustrated in this guide–the often overlooked, small, and spectacularly beautiful wildflowers so numerous in these mountains. This little *Montia* is typically 5" to 7" tall and well hidden in the grasses and other plants in the area of my search. Flowers are a crisp snow-white with olive green centers. Stamens are white, fairly prominent and topped with salmon pink anthers. If it weren't for the leaves one might think they were looking at the white version of *Claytonia obovata* (pg. 58) but while *C. obovata* has two opposite and basal leaves, this one has numerous lance to somewhat wedge-shaped leaves scattered along a tall fleshy stalk. The pencil eraser sized flowers bloom in a loose bunch at the top of the stalk. Like most members of the miner's lettuce family, the leaves are fleshy and often tinged with red. I photographed these in a dry meadow near the Packer's Peak Trailhead in late June.

LITTLELEAF MINER'S LETTUCE, SHOWY ROCK MONTIA

Montia parvifolia — 5 symmetrical petals - ovary superior

Miner's lettuce family

Like *Claytonia sibirica*, this common montia prefers moist locations. I've found it in several places but mostly on scree slopes which are wet from snowmelt, or on the edges of springs, seeps, and creeks. The stems are red or reddish green, each topped with a loose cluster of pale pink flowers, the tips of which can be either notched or entire. The basal leaves are fleshy, mostly oval, and green to olive-green, tinged with red and sometimes decorated with deep red spots. I photographed these on the open slopes above Little Duck Lake in August and shot the small photograph on the slopes above Upper Albert Lake in early September–both in the Russians.

ONE-SIDED WINTERGREEN, SIDEBELLS WINTERGREEN

Orthilia secunda — 5 symmetrical petals - ovary superior

Heath family

The stems of this wintergreen are unbranched, about 6" to 8" tall, and woody at the base. Leaves are roughly egg shaped, entire or finely toothed, and are basal–they are not stiff like little prince's pine or pipsissewa. True to its name, the small, vase-shaped greenish-white flowers bloom along one side of the upper third of the stalk–generally hanging downward. The green pistils are prominent and extend beyond the petals. Like the other wintergreens described here, this one likes deep shade and moist soils at mid to high elevations. I photographed these along the Boulder Creek Trail just above Big Boulder Lake and Boulder Creek in the Trinity Alps in August. This wildflowers is a world traveler–I have also photographed it in the Diemtigtal region of Switzerland where it is called Birngrün.

TOAD LILY
Montia chamissoi
Miner's lettuce family

LITTLELEAF MINER'S LETTUCE
Montia parvifolia
Miner's lettuce family

ONE-SIDED WINTERGREEN
Orthilia secunda
Heath family

CASCADE GRASS OF PARNASSUS - *Parnassia cirrata* var. *intermedia*
FRINGED GRASS OF PARNASSUS, ROCKY MOUNTAIN PARNASSIA - *Parnassia fimbriata*
MARSH GRASS OF PARNASSUS, BOG STAR - *Parnassia palustris*
Grass-of-Parnassus family — 5 symmetrical petals - ovary superior

Three taxa of this late-blooming wildflower are found within the range of this guide and illustrated here. These are beautiful wildflowers that inhabit fens, wet meadows, the banks of streams, and areas surrounding springs and seeps–they like their feet in water and can often be found in abundance, blooming shoulder to shoulder. This is especially true in areas along Swift Creek, around Lake Eleanor and in Bear Basin where *P. cirrata* var. *intermedia* and *P. palustris* are especially abundant. The flowers of all three species are white and prominently veined but vary in other ways. The petals of *P. cirrata* var. *intermedia* and *P. fimbriata* are narrowly elongate with a prominent fringe emerging from the base of the petals. The petals of *P. palustris* lack a fringe and are more oval–somewhat cup-shaped and with slightly pointed tips. All but *P. fimbriata* have golden yellow staminodes with long projections capped with small ball-like structures. The staminodes of *P. fimbriata* are stubby and lack the projections and ball-like structures so prominent on the other two. I would also like to note that Cascade grass-of-Parnassus is considered rare in California, restricted to the Trinity Ultramafic Sheet geologic feature of the Klamath Ranges, and appears to be strongly associated with serpentine soils. Oddly, it also occurs in Nevada and Idaho, not on serpentine but rather on alkaline soils. Perhaps these will turn out to be separate species once genetic analysis is done.

The photographs of Cascade grass-of-Parnassus were taken near Lake Eleanor just outside the Trinity Alps. Those of fringed grass-of-Parnassus were taken in the Parks Creek drainage on the Trinity Divide. Those of marsh grass-of-Parnassus were taken at L Lake in the Trinity Alps. All of the photographs were taken in September. You will also find marsh grass-of-Parnassus in the meadows around Waterdog Lake in the Russians and in the meadows around Alpine Lake in the Trinity Alps.

A note about Grass of Parnassus: This beautiful wildflower has become an absolute favorite of mine and an obsession of sorts. This is also true of wildflower enthusiasts in Scotland especially Alistair Whyte the Head of **Plantlife Scotland** who wrote this just recently:

"It's starting to feel autumnal. There's a slight chill in the air in the mornings, and up in the mountains, things are getting decidedly colder. It's great to be out and about in the hills in this weather, although it's always nice to make it back somewhere warm at the end of the day!

"It's too early to give in to the cold months yet, and our plants are making an admirable show of hanging on to the last of the summer, with species still flowering even as the days get shorter and the nights colder. We've been out in the mountains over the last month, and it's always interesting to see what plants are still in flower this late in the year. In some wet, boggy areas on the way up the hill, there's a good chance of coming across Grass of Parnassus. This lovely white flower, also known as the 'bog star', is another later-flowering species. It's always worth looking out for at this time of the year --"

I also found this by the Scots poet Andrew Lang - penned in 1882

GRASS OF PARNASSUS.

Pale star that by the lochs of Galloway,
In wet green places 'twixt the depth and height
Dost keep thine hour while Autumn ebbs away,
When now the moors have doffed the heather bright,
Grass of Parnassus, flower of my delight,
How gladly with the unpermitted bay--
Garlands not mine, and leaves that not decay--
How gladly would I twine thee if I might!

The bays are out of reach! But far below
The peaks forbidden of the Muses' Hill,
Grass of Parnassus, thy returning snow
Between September and October chill
Doth speak to me of Autumns long ago,
And these kind faces that are with me still.

CASCADE GRASS-OF-PARNASSUS
Parnassia cirrata var. *intermedia*
Grass-of-Parnassus family

FRINGED GRASS-OF-PARNASSUS
Parnassia fimbriata
Grass-of-Parnassus family

MARSH GRASS-OF-PARNASSUS
Parnassia palustris
Grass-of-Parnassus family

LEAFLESS WINTERGREEN. LEAFLESS PYROLA
Pyrola aphylla — 5 symmetrical petals - ovary superior
Heath family

Early in its life this little wintergreen is unobtrusive. Until it flowers it tends to blend in with its surroundings. It is also another species difficult to place by flower color because it is either creamy white, occasionally with pink accents, or light pink with darker pink accents. The flowering stalks are tall and brown to red and not entirely leafless–there are vestigial leaves but they are small and almost un-noticeable. The plants tend to cluster which can make them easier to spot once they begin flowering. The flowers have five petals and form a shallow downward hanging bell or bowl shape. The preferred habitat is shaded areas in the understory of mixed conifers and hardwoods. I shot the photograph of the pink version above the shoreline of Trinity Lake in mid-May and the white flowers tinged with pink along the Canyon Creek Trail in early August .

CRYPTIC WINTERGREEN
Pyrola crypta — 5 symmetrical petals - ovary superior
Heath family

I've only encountered this particular Pyrola once at the pass between Bear Basin and 7-Up Lake in the Trinity Alps in late August. It has white veining on the leaves with a naked, reddish brown stalk with tiny vestigial leaves. The nodding bell shaped flowers are creamy white or pink and bloom along the upper third of the stalk. Stamens are prominent, as is a long, curved stigma. They prefer a shaded habitat beneath mixed conifer canopies. Because of the tall and semi-leafless stalk, it could be confused with *P. aphylla*.

TOOTHED WINTERGREEN
Pyrola dentata — 5 symmetrical petals - ovary superior
Heath family

I found this little wintergreen in bloom alongside the Swift Creek Trail in mid-July. It was part of a small, scattered population growing in dry, rocky serpentine soil beneath a canopy of mixed conifer including Jeffrey pine, white fir and incense-cedar. What first attracted my attention were the bell shaped flowers which were creamy white tinged with a bit of yellow and red at the tips of the petals. The flowers emerge alternately along stalks which, in this population, were reddish brown and about 7″ tall. The deep green leaves are basal and somewhat wavy along the margins. Why is it called wintergreen? Because the leaves remain green throughout the year–long after the flowers have gone to seed and the snow falls.

WHITE-VEINED SHINLEAF, WHITE-VEINED PYROLA, WHITE-VEINED WINTERGREEN
Pyrola picta — 5 symmetrical petals - ovary superior
Heath family

This pretty little wintergreen has, true to its name, white-veined bright green, egg-shaped leaves in a basal rosette. The flowering stalk is leafless and about 8″ height with alternate, nodding white flowers with deep yellow stamens and a prominent, elongated, and curved style. As with the other wintergreens, this one likes damp, shaded sites. I photographed these in the rocky area above Lower Russian Lake in the Russian Wilderness in mid-September.

LEAFLESS WINTERGREEN
Pyrola aphylla
Heath family

CRYPTIC WINTERGREEN
Pyrola crypta
Heath family

TOOTHED WINTERGREEN
Pyrola dentata
Heath family

WHITE-VEINED SHINLEAF
Pyrola picta
Heath family

WHITE WATER BUTTERCUP - Gemeiner Wasserhahenfuss in Swiss
Ranunculus aquatilis 5 symmetrical petals - ovary superior
Buttercup family

You never know what you might find while wading around in the Trinity River on a hot afternoon. Case in point: in mid-July while I was shooting photographs for the Trinity River Restoration Project, I came across hundreds of these little beauties that look like little stars blooming at water level. They prefer slow moving water and partial shade. Each elongated stalk hosts a single snow-white flower with five petals fading to deep orange at the center. The petals surround a green, ball of multiple tiny ovaries with orange stamens radiating from beneath. Greenish-orange sepals fall away from below the petals. Leaves, if you can really call them that, look more like round, elongated green branches that radiate from the main stalks. As you can see in the habitat photograph they grow in dense floating mats. I have photographed these in Switzerland around the margins of the Lauenensee, where it is called Gemeiner Wasserhahenfuss. These photographs were taken above Steel Bridge Campground on the Trinity River.

The Trinity River beautiful and intricately tied to the majesty of the Klamath Mountains. Its massive watershed drains the southern slopes of the Scotts, the west facing slopes of the Divide, and most of the eastern portion of the Alps. It gathers water from Mt. Eddy, Scott Mountain, Thompson Peak, and Limestone Ridge as well as the north slopes of the Yolla Bolly-Middle Eel Wilderness via the South Fork before emptying into the Klamath River at Weitchepec. Streams feeding into it include the Little Trinity River, Scott Mountain Creek, Tangle Blue Swift Creek, Stewart's Fork of the Trinity River, Canyon Creek, North Fork of the Trinity River, and New River just to name a few. It pours through a variety of rock types and is home to myriad wildflowers.

WHITE WATER BUTTERCUP
Ranunculus aquatilis
Buttercup family

The Trinity River

CALIFORNIA SANDWORT

Sabulina californica

5 symmetrical petals - ovary superior

Pink family

Tiny! That's the only way to describe this delicate little sandwort. If it weren't for the fact that it blooms profusely in large populations it would be quite easy to miss-even by attentive wildflower stalkers! The stems are hair like and spindly at best making them hard to photograph even under the best of conditions. The slightest motion or just simply breathing on them can give them the jitters. The flowers are typical of the other *Sabulinas* in this guide–white with symmetrical spoon shaped petals with cream colored stamens and greenish anthers surrounding a bulbous yellow ovary. Sepals are lance shaped, yellowish. and alternate beneath the petals. Plants favor wet or dry, sandy soils in sunny, exposed sites where few other plants offer competition. The plants in these photographs were blooming in standing water and were taken in early May at Lowden Ranch a few miles below Lewiston on a flat above the Trinity River.

DOUGLAS' STITCHWORT, DOUGLAS' SANDWORT

Sabulina douglasii

5 symmetrical petals - ovary superior

Pink family

Here is another Scott Mountain wildflower I found on one of my many visits to this botanically rich site. The tiny white flowers are anywhere from 3/16" to just over 1/4" in diameter and arranged in a loosely branched formation on thin, hair-like stalks. They are so light that photographing them presents a real challenge–just breathing set them into wild motion. Flower petals are striped lengthwise, gathering around an olive green center, a prominent yellowish ovary, and several gently curved stamens. Leaves seem to be almost non-existent. Because of the scattered nature of the populations this is an easy one to pass by so pay close attention as you walk and watch your step!

NUTTALL SANDWORT

Sabulina nuttallii var. *gregaria*

5 symmetrical petals - ovary superior

Pink family

Look for this tiny wildflower on dry hillsides at higher elevations, mainly on the serpentine soils of the Red Trinities and the Trinity Divide. It forms mats ranging in height from 3" to 5" which can cover several square feet. The stems and awl shaped leaves are a rusty-olive green, hairy, and sticky-glandular. The tiny white flowers are five-petaled with five stamens and five sepals. They bloom in loose clusters. I photographed this bunch near the top of the pass between Deer and Summit Lakes in the Trinity Alps in early September. Another place where this sandwort is especially showy is at the pass between Union Meadows and Battle Canyon in the Trinity Alps–usually in June after the snow has melted.

CALIFORNIA SANDWORT
Sabulina californica
Pink family

DOUGLAS'STITCHWORT, DOUGLAS' SANDWORT
Sabulina douglasii
Pink family

Cement Bluff Lake: Located on the Trinity divide in a basin filled with wildflowers including Nuttall sandwort.

NUTTALL SANDWORT
Sabulina nuttallii var. *gregaria*
Pink family

BELL CATCHFLY, CAMPANULATE CATCHFLY, RED MOUNTAIN CATCHFLY

Silene campanulata — 5 symmetrical petals - ovary superior

Pink family

This wildflower is common and widespread throughout the Klamath Mountains at lower elevations. It likes shaded openings in forested areas and often grows in large groups. Stems are erect with nodding flowers. Leaves are more or less oval to lance shaped. The white to greenish-white flowers have five petals, each of which is deeply divided into four to eight linear segments. The reflexed petals sit deep in the bell shaped calyx, which gives rise to its common name. I have photographed these catchflys along the lower sections of the trails up Canyon Creek, the North Fork of the Trinity River, Stewart's Fork, around Trinity and Shasta Lakes. These photographs were taken above Canyon Creek in June.

DOUGLAS' CATCHFLY, DOUGLAS' CAMPION, MOUNTAIN CAMPION

Silene douglasii var. *douglasii* — 5 symmetrical petals - ovary superior

Pink family

Generally, you will find this pretty little catchfly blooming in large troops in open, rocky and meadowy areas at mid to high elevations. It also seems fond of steep and almost inaccessible ridgetops. Look for the green and deep purple striped "bell" with the two-deeply lobed, white petals folding back from the opening. The plants in these photographs were about 12" to 14" tall. I shot this photograph on the ridge above Big Blue in the Russians in September. It was abundant all along the ridge running north from Upper Albert Lake to Taylor Peak.

LEMMON'S CATCHFLY, LEMMON'S CAMPION, PINE PINK

Silene lemmonii — 5 symmetrical petals - ovary superior

Pink family

I found this little catchfly blooming on the open, rocky slopes between Lower and Middle Caribou Lakes in the Trinity Alps in late August, but it is fairly common at mid to high elevations throughout the area covered by this book. The elongated calyx forms a bell with alternating light and dark green stripes. Thin, curly white four-lobed petals emerge from the bell and flare outward. The long hair-like things without anthers are the prominent styles. The entire plant rarely exceeds 10" in height.

MENZIES' CAMPION

Silene menziesii — 5 symmetrical petals - ovary superior

Pink family

It pays to have friends who are committed botanists, professional or otherwise. In this instance it was Joaquin Hale who found this pretty little campion and who was gracious enough to show me the population he had found. The small flowering plants were tucked into a semi-shaded area beneath a canopy of streamside trees and shrubs and would have probably gone un-noticed by the casual passer-by. Plants are 3" to 5" in height with stalks and opposite egg-shaped leaves covered with short, soft hairs. Flowering stalks rise from leaf axils, each with a single flower. Flowers are white with deeply cleft petals rounded at the tips. Flower centers are green with prominent stamens and creamy yellow anthers. The entire plant presents a delicate and fragile aspect. I photographed these in the Lewiston Basin in late May.

BELL CATCHFLY
Silene campanulata
Pink family

DOUGLAS' CATCHFLY
Silene douglasii var. *douglasii*
Pink family

LEMMON'S CATCHFLY
Silene lemmonii
Pink family

MENZIES' CAMPION
Silene menziesii
Pink family

STRINGFLOWER
Silene nelsonii
Pink family

5 symmetrical petals - ovary superior

This is a gorgeous wildflower. It is showy and large enough to be noticed by even the most casual observer. Its five snow-white petals are cleft into fou long narrow lobes, the central two being wider and longer than the outer two. The entire plant grows to about 5" in height. Leaves are small, grayish-green, basal, and covered with fine hairs. I photographed these at Natural Bridge near Hayfork and have never found it any place else but I understand it is not uncommon in Trinity County. Alice Jones reported finding it years ago in Norwegian Meadows (private property) near Trinity Center. Julie Kierstead has seen it at Burnt Ranch, on Underwood Mountain, and on the Yolla Bolla Ranger District Compound.

BOUNCING BET, SOAPWORT
Saponaria officinalis
Pink family

5 symmetrical petals ovary superior

This is an invasive naturalized wildflower found along the Trinity River where it intermingles with stands of willow and other plants at the waters edge. It also does well along roadsides and in oak woodlands where the ground has been disturbed. The plants I've photographed have been 24" tp 36" tall, leggy, and loosely branched with loose clusters of white flowers that age to pink. The flower petals are slightly reflexed, narrow at the center flaring outward, and terminating in rounded and notched tips. It's hard to miss this one if you spend any time at all on the Trinity River. I photographed these near Junction City in August.

STRINGFLOWER
Silene nelsonii
Pink family

BOUNCING BET, SOAPWORT
Saponaria officinalis
Pink family

Thunderstorm over the Upper Trinity River

RANGER'S BUTTONS, SWAMP WHITEHEADS, WOOLY-HEAD PARSNIP
Angelica capitellata 5 symmetrical petals - ovary inferior
Carrot family

This gorgeous wildflower prefers wet places such as the shorelines of lakes, streams, or marshy meadows. It grows to 5′ or more in height on stout, branched stems. The leaves are oblong, 4″ to 15″ and divided three or four times, with each leaflet lance shaped, irregularly toothed, or lobed and 1/2″ to 5″ long. The white flowers, aging to light brown, are distinctive, growing in large clusters of tight, round heads at the tops of the stems. I photographed these in a wet meadow below Mt. Ashland. And, by the way, I retired from the Forest Service after 35 years with the agency and have to admit that on my spiffiest uniforms I never had buttons like these! **Note:** While there are no recorded sightings in the Trinity Alps or Russians, there are several from along the Shackleford Creek drainage in the Marble Mountains.

ELK'S CLOVER, CALIFORNIA SPIKENARD, PRAIRIE SAGEWORT
Aralia californica 5 symmetrical petals - ovary inferior
Ginseng family

Elk clover is found throughout the area at low to mid elevations where it is associated with damp, shaded sites alongside streams, near seeps, springs, and the shorelines of lakes and ponds. It's not a clover at all and the flowers don't even resemble clover flowers. They are striking, however, and these tall perennials are covered with beautiful white globes in the spring. The flowers are followed by hanging bunches of deep purple fruit, which bears seem to enjoy. I photographed the flowers along Canyon Creek in mid-July, and the fruit at the Stewart's Fork Trailhead in early November.

MOUNTAIN BOYKINIA
Boykinia major 5 symmetrical petals - ovary inferior
Saxifrage family

This wildflower inhabits streambanks, springs, and other damp, shaded areas where it often grows in dense colonies. Some people mistakenly call it elephant ears because of slight similarities in leaf size and structure but elephant ears are much larger. Overall height of the plants can range from 2′ to 3′, with deeply cleft and toothed leaves ranging from 4″ to 8″ across. The small white flowers grow in dense clusters with petals about 1/4″ long and are decorated with bright yellow nectar glands in the centers which attract a variety of insects. I photographed this one at a spring along the trail to Stonewall Pass in the Trinity Alps in May. When fall sets into the high country the leaves put on quite a show turning beautiful shades of gold and red.

RANGER'S BUTTONS
SWAMP WHITEHEADS
Angelica capitellata
Carrot family

ELK'S CLOVER
CALIFORNIA SPIKENARD
Aralia californica
Ginseng family

MOUNTAIN BOYKINIA
Boykinia major
Saxifrage family

BASTARD TOADFLAX, CALIFORNIA BASTARD TOADFLAX
Comandra umbellata — Alternate leaves - 5 symmetrical petals ovary superior
Comandra family

There is a trail above Shasta Lake that I have visited for years and along that trail I've always been intrigued by a small plant growing there. I've always looked for any indication that it produces flowers and during all that time–nothing. That changed recently when, to my surprise, there they were! Hundreds of flowers borne in small clusters at the tips of 6 to10" stems exhibiting the habit of a ground cover. The tiny flowers lack petals but the 4 to 5 white sepals are flower-like and surround both male and female structures. The leaves alternate on the stems–long and sharply pointed at the tips. The plant is hemiparasitic in that they produce chlorophyll but get some nutrients from other plants. CalFlora indicates numerous sightings along the entire length of the Trinity River.

COW PARSNIP, GIANT HOGWEED
Heracleum maximum — 5 symmetrical petals - ovary inferior
Carrot family

The most noticeable thing about cow parsnip is its size. I've seen plants over 7' tall towering over everything else around. It hangs over trails and dusts passers-by with loose flower parts all summer long. When the stems are broken the plant has a distinctive odor, which I find surprisingly pleasant–but I've heard others complain about it. Look for them in any mid-to-high elevation meadow and alongside trails where you will, no doubt, do battle with it while carrying a heavy pack. I've photographed this from the coast to the upper end of Siligo Meadows in the Trinity Alps. This photograph was taken along the East Fork of Stewart's Fork Trail below Bee Tree Gap in mid-August.

CREVICE ALUMROOT, SMALL-FLOWERED ALUMROOT
Heuchera micrantha — 5 symmetrical petals - ovary inferior
Saxifrage family

This is a beautiful and graceful plant that loves moisture and shade. It is most often found growing in rocky outcrops in the proximity of streams or deep, shaded canyons. It is gregarious,–often occurring in large colonies. The tiny creamy-white flowers bloom in a tall, almost pyramidal inflorescence. Flowering stalks in deep shade are green but if they get more sun they can be reddish. The oval shaped leaves are lobed and usually covered with tiny glanular hairs. I photographed this bunch in full flower at the Little Deep Creek Crossing along Stewart's Fork Trail in May. **Note:** It is an important garden plant is some areas.

GRAY'S LOVAGE
Ligusticum grayi — 5 symmetrical petals - ovary inferior
Carrot family

This is a common wildflower found in drier, open meadowy areas and alongside trails at low- to mid-elevations. Tall, leafless stems rise out of a large basal clump of deep green, feathery leaves. The inflorescence is called a compound umbel, i.e. an umbel of smaller umbels, in which smaller branches radiate from the same point at the top of the main stem. Each umbel branch is topped with another umbel, each terminating in a spray of tiny, elegant snow-white flowers. Small but prominent stamens radiate away from the center of each flower. I photographed these along the Boulder Creek Trail in August.

BASTARD TOADFLAX
Comandra umbellata subsp. *californica*
Comandra family

COW PARSNIP, GIANT HOGWEED
Heracleum maximum
Carrot family

CREVICE ALUMROOT
Heuchera micrantha
Saxifrage family

GRAY'S LOVAGE
Ligusticum grayi
Carrot family

BIGSEED BISCUITROOT, LARGE-FRUITED LOMATIUM
Lomatium macrocarpum 5 symmetrical petals - ovary inferior
Carrot family

My encounters with this wildflower have almost always been at higher elevations on rocky ridgetops and open slopes. The flowers are a soft purplish-white (sometimes yellow) and gathered in dense compound umbels at the tops of short stems. The somewhat fleshy leaves are a gray-green and finely dissected. Chipmunks are especially attracted to the flower heads in late summer much as they are to the dry flower heads of the various buckwheats. They harvest the seeds and store them away for the long winters ahead. I shot this photograph on the ridge between Foster and Lion Lakes in the Trinity Alps in early June.

SIERRA SAXIFRAGE
Micranthes aprica 5 symmetrical petals - ovary inferior
Saxifrage family

This is one of those "up close and personal" wildflowers that requires you to get down on your hands and knees. If you do you will be treated to a view of the bright white flowers, deep orange anthers, and the bright green pistils. It's when you are nose to nose with this saxifrage that you actually see the colorful structure. The plants are small–the entire flowering cluster only about 1½" wide and the stems about 4" - 5" tall. Leaves are mostly oval but deeply cleft. It favors rocky, wet alpine meadows next to receding snowbanks or meltwater streams. I photographed these in a small meadow below Kangaroo Lake on the Trinity Divide and the inset in Union Meadows in the Trinity Alps in early June.

BRITTLE-LEAVED SAXIFRAGE, FLESHY-LEAVED SAXIFRAGE
Micranthes fragosa 5 symmetrical petals - ovary inferior
Saxifrage family

This is a saxifrage that prefers open, damp, grassy meadows. It grows to about 7" tall with loosely branched and slightly hairy stems. Clusters of white flowers bloom at the tops of each stem. The petals flare away from the center of the flower and several stamens topped with buttery yellow anthers surround the central pistil. I photographed these in and amongst many others in Doe Flat in the Trinity Alps in mid-August. I have also found them in the meadow surrounding 7Up Lake, also in the Trinty Alps.

OREGON SAXIFRAGE, EARLY SAXIFRAGE
Micranthes oregana 5 symmetrical petals - ovary inferior
Saxifrage family

This pretty little saxifrage has a basal rosette of egg-shaped leaves toothed along the margins. The plants can grow to 14" with fairly erect, branched stems. Clusters of small white flowers crowd near the tops. Flower petals are nearly round to egg-shaped and white. The anthers are orange to purplish. It favors wet meadows, and the shaded edges of lakes and streams. I found and photographed these around a spring just above the shoreline of Sugarpine Lake in the Trinity Alps in mid-September.

BIGSEED BISCUITROOT
Lomatium macrocarpum
Carrot family

SIERRA SAXIFRAGE
Micranthes aprica
Saxifrage family

BRITTLE-LEAVED SAXIFRAGE
Micranthes fragosa
Saxifrage family

OREGON SAXIFRAGE, EARLY SAXIFRAGE
Micranthes oregana
Saxifrage family

SWEET-CICELY, MOUNTAIN SWEET CICELY
Osmorhiza berteroi
Carrot family
5 symmetrical petals - ovary inferior

This is a mid-elevation wildflower you will find in damp, deeply shaded sites. The trifoliate leaves are mostly basal, have serrate margins, and occur in opposite pairs on long stalks. Flowers are small, white to cream, and bloom in a loose cluster or compound umbel at the tops of long stems. The plants can grow to a height of 2' or more. Native Americans in several areas ate the roots and used parts of the plant as a medicine for coughs and colds. I photographed these along Stewart's Fork Trail above Morris Meadows in mid-June.

THREE-PARTED MITERWORT, THREE TOOTHED MITERWORT, THREE-CORNERED BISHOP'S CAP
Ozomelis trifida
Saxifrage family
5 symmetrical petals - ovary inferior

The tiny, but beautiful snow white flowers are striking once you get up close and personal with them. The flowering stalks are over 12" tall, rising out of widely oval, and palmately lobed leaves. The flowers bloom all along the stem and are cup shaped with petals that flare outward and are deeply cleft. I understand they can also be a light purple. They seem to prefer moist, shaded places. I found this beautiful saxifrage blooming along the lower reaches of Stewart's Fork Trail in late May a couple of years ago. I can't tell you how many times I've walked that trail but in the course of 63+ years let's just say "many" This was the first time I'd ever seen it in bloom. All this leads to my assertion that you can never be overly familiar with an area and what happens there when you're not looking.

JOINT-LEAVED SAXIFRAGE
Saxifragopsis fragarioides
Saxifrage family
5 symmetrical petals - ovary inferior

The flowers of this gorgeous saxifrage are multi-colored, with narrow white petals and prominent stamens accented with bright yellow and red ovaries. They bloom on hairy reddish-tan upright stalks (some approaching 6"). Leaves are oval, dark green, sharply toothed, and most slightly folded at the mid-rib. I photographed these after encountering dozens of them while climbing around the granite outcrops below Upper Albert Lake in the Russians in late August. Most of the little saxifrage species shown in this guide are colorful but this one takes the prize.

WHIPPLEVINE, YERBA DE SELVA
Whipplea modesta
Hydrangea family
5 symmetrical petals - ovary inferior

Classified as a sub-shrub, this pretty little wildflower blooms at the ends of decumbent stems with brownish bark that peels away in small, narrow strips. It tends to grow in large, spreading populations and is not particular about habitat–it can be found in chaparral, open forest, as well as along streambanks. Four to six white flowers form a dense raceme each having four to six petals with several prominent stamens. I photographed this in Prairie Creek State Park on the coast in late April. It is widespread throughout the low to mid elevations of the Klamath Mountains and Coast Range.

SWEET-CICELY
Osmorhiza berteroi
Carrot family

THREE-PARTED MITERWORT
Ozomelis trifida
Saxifrage family

JOINT-LEAVED SAXIFRAGE
Saxifragopsis fragarioides
Saxifrage family

WHIPPLEVINE, YERBA DE SELVA
Whipplea modesta
Hydrangea family

TRINITY PENSTEMON, TRACY'S BEARDTONGUE
Penstemon tracyi
Plantain family

5 irregular petals forming a tube

This beautiful penstemon is a rare Klamath Mountains endemic confined to metamorphic rock crevices and cliff faces in the Trinity Alps. The pink and white flowers form elongated tubes arranged in drooping clusters and are hairy on the lower lip. Stems are woody. The leaves are mostly basal, oval to rounded, thick. and leathery. Smaller leaves are arranged in opposite pairs up the flowering stalk. It took two years and several trips up Packer's Peak to find them in bloom but I finally lucked out when I found it in rocky outcrops halfway up the trail from Big Flat. Though a bit difficult to get to, it was worth the effort. I photographed these in early June.

CALIFORNIA SKULLCAP
Scutellaria californica

5 irregular petals forming a tube

CALIFORNIA SKULLCAP - NOSE SKULLCAP HYBRID
Scutellaria californica X Scutellaria antirrhinoides
Mint family

California skullcap is mostly white to light yellow and can be tinged with pink or blue. It hybridizes readily with nose skullcap (see page 358) producing beautiful white and purple flowers. The flowers are slightly hairy beneath. Leaves of both are mostly oval to wedge shaped with rounded tips. I found both of these skullcaps blooming shoulder to shoulder along the trail to Canyon Creek Lakes in pockets of an exposed granite outcrop where I've photographed several other wildflowers through the years. I had seen the plants before but this was my first sighting of either of them in bloom. There were significant populations of both and almost impossible to miss although as I was shooting them several backpackers passed by totally obivious to them until, when asked what I was doing, I pointed them out.

TRINITY PENSTEMON, TRACY'S BEARDTONGUE
Penstemon tracyi
Plantain family

CALIFORNIA SKULLCAP & CALIFORNIA-NOSE SKULLCAP HYBRID
Scutellaria californica & Scutellaria californica X antirrhinoides
Mint family

TRINITY MONKEYFLOWER
Erythranthe trinitiensis
Lopseed family

5 irregular petals forming a tube

This small and showy monkeyflower was described in 2013 and, except for this guide, won't be found in any of the current botanical manuals. It is diminutive; most plants are less than 2" tall and the flowers pea-sized. It appears to be endemic to the serpentine soils found throughout this area and seems to prefer wet meadows and somewhat open ground beneath a canopy of mixed conifers. The floral tube is a deep buttery yellow becoming white as the petals open and limned with rose red at the petal tips. It is indeed, striking. I found a particularly nice population interspersed with Siskiyou phacelia in a high elevation area on the west side of the Trinity Divide in late June. I'll wager odds that even seasoned botanists have been walking over this one for years without noticing it.

COILED-BEAK LOUSEWORT
Pedicularis contorta
Broomrape family

5 irregular petals forming a tube

This odd little wildflower is fairly common in this area. It is readily identifiable because of the small, drooping white flowers with pale yellow accents punctuated by deep maroon spots. Flowers resembling little louse critters is one of the reasons for the name the other is that centuries ago louseworts were thought to repel lice (it was common practice at one time to associate plants that looked like animals with certain medical and/or cultural properties). Basal and lower stem leaves are 2" to 8" long, linear and highly divided into linear segments. Flowers occur along a tall 10" to 12" spike. The upper lip curves sharply downward to meet the three lower lobes. I photographed these in mid-August in the meadows at the back of Upper Caribou Lake. I've found it in the drier meadows surrounding Carter's Little Mud Lake and Caribou Lake in the Trinity Alps, below the outlet of Big Blue in the Russian Wilderness, and around Wolverine and Upper Wright Lakes in the Marbles.

HOT ROCK PENSTEMON
Penstemon deustus
Plantain family

5 irregular petals forming a tube

This penstemon, as its name implies, likes dry, open, rocky outcrops, and hillsides. The flowers bloom in multiple whorls along a stem that can reach anywhere from 10 to 20 inches tall. The petals are shallowly recurved and white (sometimes tinted with faded magenta) and streaked with deep purple. The buttery-yellow flower tubes are small, about 1/2" long. The yellowish-green leaves are opposite and raggedly toothed. I have found it at lower elevations along the Trinity River near Reading Creek and at the tops of ridges in mid-summer and early fall throughout the Trinity Alps and Russian Wilderness Areas. I photographed these near Middle Falls along Canyon Creek in mid-May.

TRINITY MONKEYFLOWER
Erythranthe trinitiensis
Lopseed family

COILED-BEAK LOUSEWORT
Pedicularis contorta
Broomrape family

HOT ROCK PENSTEMON
Penstemon deustus
Plantain family

BEAUTIFUL STICKSEED
Hackelia bella
Borage family

5 symmetrical petals forming a tube

This is one of the showiest stickseeds to bloom in this area and one that few hikers could possibly miss. It is widespread throughout the mid-elevations where it grows in large colonies at the edges of meadows where soils are dry. The plants are loosely branched, often reaching heights of 2′ or more, with open clusters of snow white flowers that are tinged with pink (see inset). I photographed these in Union Meadows in the Trinity Alps in late May where it offers up a great trail-side display.

DRAPERIA, WEIRD FLOWER
Draperia systyla
Waterleaf family

5 symmetrical petals formning a tube

How or why people give certain names to some wildflwowers baffles me sometimes and "weird" flower is one of those. I have to admit that I find nothing particular weird about it. It's a beautiful California endemic that is found from the Trinity Alps south through the Sierra Nevada. It's small with several stems about 5″ long. Leaves are simple, opposite, hairy, and 1′ to 2″ long. The funnel shaped flowers range from white to pale violet, about ½″ long and custered at the ends of short branched stems. Preferred habitat includes woodlands, talus, and rock crevices from low to mid-elevations. I shot this photograph above Shasta Lake but have found it in abundance along the Stewarts Fork Trail in the Alps.

DWARF HESPEROCHIRON
Hesperochiron pumilus
Borage family

5 symmetrical petals joined into a tube

The beautiful little flowers dotted a wet, grassy meadow near where Alice Jones and Bill Ferlatte recorded their finds in the 1950s and 60s. As my friend Max Creasy likes to say–these are "belly flowers" and in order to fully appreciate them you need to get down in the dirt for a look. They are thumbnail size, with a corolla that is disc to slightly cupped in shape standing on a short stalk about 1½″ tall above a rosette of oval to lance shaped leaves. The petals are bright white with deep purple veins emanating from an orange-yellow center arrayed with prominent yellow stamens and dark purple anthers. The center of the flower is hairy. It took a couple of years of concerted searching but I finally found this tiny and reclusive wildflower in full bloom. Documented sightings put the flowering time in mid-June but that was a total bust. Frustrated, I finally began searching in early May and was rewarded when I found a small, scattered population. The old saying holds true: Right place–right time! I photographed these in Big Flat in the heart of the Trinity Alps.

SCOTT MOUNTAIN PHACELIA
Howellanthus dalesianus
Waterleaf family

5 symmetrical petals forming a tube

This diminutive phacelia is a rare ground hugger and hard to spot. It is endemic to serpentine soils and prefers areas that have been disturbed alongside trails and roads. The flowers grow at the ends of short pedicels and are about ¼″ across, creamy white, with deep purple spots near the base of each petal. Inspect the flowers closely and you will see that the petals are united at their bases and form a broad bowl-shaped tube. The hairy and deeply veined leaves vary from 1″ to about 2″ in length with smooth edges. I photographed these on Scott Mountain in a year when they were abundant, but most years it takes some looking to find them. A good place to see them is along the Fen Trail at Kangaroo Lake where they are also described in the interpretive sign.

BEAUTIFUL STICKSEED
Hackelia bella
Borage family

DRAPERIA, WEIRD FLOWER
Draperia systyla
Borage family

DWARF HESPEROCHIRON
Hesperochiron pumilus
Borage family

SCOTT MOUNTAIN PHACELIA
Howellanthus dalesianus
Waterleaf family

WHITE WATERLEAF, FENDLER'S WATERLEAF

Hydrophyllum fendleri var. *albifrons* 5 symmetrical petals forming a tube

Waterleaf family

Plants are about 12" in height, with bright green leaves 6" to 8" long and deeply cleft. Leaves lack the water spots–or the whitish discolorations typical of California waterleaf. The snow-white flowers bloom in a compact head, all but hidden beneath the canopy of leaves. This waterleaf is a bit secretive–hiding in deep shade amongst other wet meadow plants. Finding it can be an accidental affair unless, of course, you are spending a lot of time on your hands and knees crawling through dense undergrowth looking for it. On the advice of Julie Knorr, I found a nice population growing along the trail to South Fork Lakes in mid-June. California waterleaf (page 364) is also in the area and could be confused with white waterleaf, the two species often rub shoulders.

NUTTALL'S LINANTHUS

Leptosiphon nuttallii subsp. *nuttallii* 5 symmetrical petals forming a tube

Phlox family

The five-petaled flowers are a creamy white with deep yellow anthers and quite showy. The leaflets are tough, deeply palmately divided, and almost spiny. The plants are 7" to 10" tall but are social and hard to miss when in bloom. I shot these, with my grandson, along Boulder Creek just before starting the climb to Conway Lake in August.

BOGBEAN, BUCKBEAN, ALPINE FRINGE

Menyanthes trifoliata 5 symmetrical petals forming a tube

Buckbean family

To see this species you have to get your feet wet. Leaf and flowering stalks emerge from submerged rhizomes. Leaves are alternate with three obovate leaflets. The flowering stalks terminate in a striking raceme of hairy white and pink tinged five petaled flowers. Flowers are tubular with petals spreading away from the center. The yellowish-green style is elongated and taller than the stamens wich are topped with reddish-brown anthers. The entire plant can reach to 10 or 12". These photographs were taken at Cedar Lake in late June and the photograph of the leaves provided by Julie Kierstead. **Note:** The leaves have sometimes been used as a hops substitute in the beer making process. **On another note:** Buckbean is worldly in its distribution. I have photographed it at Lauenensee in Switzerland where it is called Fieberklee (Fire pea). **One more note:** My Grand-daughters thought the common names for this wildflower rather un-appealing and decided that it should be called "Alpine fringe" and as Julie Kierstead notes, common names are a dime a dozen–you can call a plant anything you want. I've chosen Alpine fringe in hopes it inspires more than my youngest generation of wildflower enthusiasts!

INTERWOVEN NAVARRETIA, NEEDLE-LEAVED NAVARRETIA

Navarretia intertexta 5 symmetrical petals forming a tube

Phlox family

This tiny navarretia grows to 2" to 4" tall with needle-like leaves and tubular shaped flowers massed in a ball at the tops of the stems. The flowers range from white to pale blue. The petals fan outward exposing yellow anthers and long stamens. I've found it on dry and exposed hillsides which appears to be its preferred habitat. It is fairly widespread in the Trinities and Marble Mountains. I photographed this one in Big Flat in the Trinity Alps but have also seen it on Musser Hill near Weaverville and along Canyon Creek Road.

WHITE WATERLEAF, FENDLER'S WATERLEAF
Hydrophyllum fendleri var. *albifrons*
Waterleaf family

NUTTALL'S LINANTHUS
Leptosiphon nuttallii subsp. *nuttallii*
Phlox family

BUCKBEAN, BOGBEAN
Menyanthes trifoliata
Buckbean family

INTERWOVEN NAVARRETIA
Navarretia intertexta
Phlox family

CANYON NEMOPHILA, WHITE NEMOPHILA
Nemophila heterophylla — 5 symmetrical petals forming a tube
Waterleaf family

This little nemophila (5″ to 8″ tall) favors partially to deeply shaded slopes and flats and can cover large areas like a carpet with masses of flowers. It is usually associated with an oak or mixed hardwood canopy. The delicate white, five petaled flowers, the fuzzy oak-shaped leaves, and the area in which they grow are distinctive. The anthers are white with a tiny purple stripe running the length and they extend well beyond the cup surrounding the ovary. I shot the large photograph on Bowerman Ridge and the small inset above French Gulch on the Trinity Divide, both in June.

BABY BLUE EYES
Nemophila menziesii var. *atomaria* — 5 symmetrical petals forming a tube
Waterleaf family

This is, by far, the largest and showiest of the local *Nemophilas*. The plants are between 4″ and 6″ in height and olive green in color with hairy leaves that are opposite and pinnately lobed. The flowers are large (3/4″ to 1″ in diameter) and variable in terms of color. Some are snow-white and faintly tinged with blue at the margins of the petals. Others are a vibrant bluish-purple except for a white center. Both have lines of small black or deep purple dots extending from the tube to the tips. I found huge colonies of both color variations blooming in conjunction with cream cups and heron's bill on open, dry, rocky slopes along Canyon Creek Road in mid-April. The drive up Canyon Creek is beautiful at any time but doubly so when these are blooming.

LITTLEFOOT NEMOPHILA, MEADOW NEMOPHILA
Nemophila pedunculata — 5 symmetrical petals forming a tube
Waterleaf family

Several varieties of *Nemophila* bloom throughout the lower elevations of these mountains. All of them seek out the same kind of habitat–semi-shaded and moist forested or grassy areas. The flowers of this one are small, white, and sometimes blue with deep bluish-purple dotted veins eminating from the light green centers and running outward to the petal tips. The leaves are deeply cleft and fuzzy. Other occurances include the Upper Trinity River, the Trinity Divide, as well as in the Russians. I photographed these along Deadwood Road near Lewiston in May.

SIERRA BABY BLUE EYES, SIERRA NEMOPHILA
Nemophila spatulata — 5 symmetrical petals forming a tube
Waterleaf family

Purple spots at the tips of each petal distinguishes this delicate *Nemophila* from its cousin the meadow nemophila which is white. Their habitat requirements are similar as this one also prefers semi-shaded and damp grassy areas. Both are low growing. The flowers are relatively short lived–disappearing quickly as daily temperatures build in the spring. It is common around the north end of the Sacramento Valley in the foothills between Redding and Shasta Lake but I've not encountered them anywhere else on the Trinity side of the Divide other than the sighting I made in Hay Gulch above Trinity Lake which is where I photographed these.

CANYON NEMOPHILA, WHITE NEMOPHILA
Nemophila heterophylla
Waterleaf family

BABY BLUE EYES
Nemophila menziesii var. *atomaria*
Waterleaf family

LITTLEFOOT NEMOPHILA
Nemophila pedunculata
Waterleaf family

SIERRA BABY BLUE EYES,
Nemophila spatulata
Waterleaf family

SERPENTINE PHACELIA
Phacelia corymbosa
Waterleaf family

5 symmetrical petals forming a tube

The most common feature of the phacelias represented in this guide is the way the blooms unfurl from tight coils along the flowering stems, giving rise to some of the common names like scorpion weed and scorpion plant. The younger plants are a soft green in their entirety (except for the flowers of course) while older, established plants tend to have red- or reddish-green stems and red tinted leaves. The flowers are white and cup shaped with stamens extending well beyond the cup. Preferred habitat is on exposed, dry, rocky slopes–occurring on, but not limited to, serpentine soils. I photographed these along the Upper Trinity River near Eagle Creek but I have also found them on the Stony Ridge Trail in the Trinity Alps. They can also be found on the north facing slopes of Scott Mountain along Highway 3.

ROCK PHACELIA, KAWEAH RIVER PHACELIA
Phacelia egena
Waterleaf family

5 symmetrical petals forming a tube

This is a fairly common phacelia found throughout the foothills surrounding the Central Valley. Within range of this guide it has been found mainly along the Trinity and New River drainages. This is a gregarious plant, often found in large populations. Its habit is mostly upright with stiff, hairy stems that are spreading and form a mound with the leaves. The lanceolate leaves are stiff and mostly basal with prominent veins. Flowers are creamy white and bell-shaped with long oblong stamens. I took these photographs alongside Highway 299 on the west side of Oregon Mountain and along Canyon Creek road.

VARIABLE-LEAF SCORPIONWEED, VIRGATE PHACELIA, VARIED-LEAF PHACELIA
Phacelia heterophylla var. *virgata*
Waterleaf family

5 symmetrical petals forming a tube

This is one of the large phacelias found in these mountains. It is strongly erect with multi-branched stems. The white, cupped flowers unfurl from tight bright green hairy coils. The flowering stalk is stiff and hairy and the leaves are mostly basal and prominently veined–smaller leaves are scattered upward along the stalk. Some of the plants I found along East Shore Drive (Trinity Mountain Road) were approaching 3′ in height though 18″ to 24″ was the norm. The population along East Shore Drive was in full bloom on the hottest and driest road shoulders, cutbanks, and other disturbed areas. It appears that it is not limited to a particular substrate as I've found it in shale and serpentine.

SEREPNTINE PHACELIA
Phacelia corymbosa
Waterleaf family

ROCK PHACELIA
Phacelia egena
Waterleaf family

VARIABLE-LEAF SCORPIONWEED
Phacelia heterophylla var. *virgata*
Waterleaf family

CHANGEABLE PHACELIA, CATERPILLAR PLANT

Phacelia mutabilis 5 symmetrical petals forming a tube
Waterleaf family

This phacelia is common throughout the subalpine areas where it loves the dry, sandy, and disturbed soil alongside trails where it can be quite abundant. As is typical with most phacelias, the leaves are a dead give-away. They are decidedly lance-shaped, deeply veined and covered with stiff hairs. The white flowers have long and pronounced stamens topped with orange to red anthers. The flowers occur in thick, coiled clusters that unfurl as they mature. I photographed this plant on the trail between Doe and Little Granite Lakes in the Trinity Alps in mid-august.

TALL PHACELIA

Phacelia procera 5 symmetrical petals forming a tube
Waterleaf family

Many times, after making a new wildflower discovery I am so intent with setting up the camera and its settings that I temporarily overlook the things around me. As it invariably happens, while the camera is doing its thing, I have a few moments to look around before adjusting settings for different exposures. It was during one of these moments that I noticed something entirely new right there in plain sight. As it turns out it was a population of tall phacelia–a species I'd been relentlessly and unsuccessfully trying to find.

As far as size goes tall phacelia comes by its name honestly–I've never seen a taller phacelia. Some of the plants in this population were 3′ to 5′ tall. The creamy white flowers unfurl in typical phacelia fashion but these are much larger and more cup shaped with the petal tips flaring away from the tube exposing elongated sepals and anthers. I photographed these near the outlet of Upper Wright Lake in the Marbles in July.

CHANGEABLE PHACELIA
Phacelia mutabilis
Waterleaf family

TALL PHACELIA
Phacelia procera
Waterleaf family

POPCORN FLOWER & CRYPTANTHA
Plagiobothrys species and *Cryptantha* species — 5 symmetrical petals forming a tube
Borage family

This is one group of wildflowers that can cause even seasoned botanists no end of heartburn. There are over 100 species in California alone so, to simplify things we have chosen to lump them into one big hairball. If they resemble the photographs here you can be relatively certain you have found what we refer to as "popcorn flowers." Generally speaking the white flowers of both bloom in clusters, are small, have ruffled edges, with little yellow, orange, or white decorations surrounding the throat. They can range from a couple of inches in height to over 6." *Cryptanthas* tend to be bristly-hairy while *Plagiobothrys* tend to be softly hairy. As a general rule, *Cryptanthas* prefer dry, exposed places while *Plagiobothrys* prefer areas that are damp or wet. They are found throughout the range of this book.

CALIFORNIA MISTMAIDEN, CALIFORNIA ROMANZOFFIA
Romanzoffia californica — 5 symmetrical petals forming a tube
Waterleaf family

There are steep, wet cliff faces along Canyon Creek Road where these bloom so thickly that they look like elegant white waterfalls cascading over the rocks. For that reason alone this little wildflower is appropriately named. The tiny, delicate, funnel or bell shaped white flowers bloom on loosely branched, slender stems. They are lightly veined with a deep yellow throat. The small, deep green leaves are basal, hug the ground, and are mostly round but deeply cleft. The plants like running water or other wet or marshy places. I photographed these along Canyon Creek Road in mid-April.

CREAM STONECROP
Sedum oregonense — 5 symmetrical petals forming a tube
Stonecrop family

Like several other species of wildflowers noted in this book, telling the difference between those that are closely related can be confusing for the amateur. Cream stonecrop and the other similar white to yellow flowered stonecrops are no exception. The flowers are white as they emerge, turning yellowish-white as they mature. Petals are sharply pointed and flare away from the ovaries which mature a brilliant red. Sepals are yellowish-green. Yellow anthers are prominent. Basal leaves are gray green, fleshy, and lined with maroon accents. They are known from the Trinity Alps and Marble Mountains and north to Mount Hood. I photographed these in early June.

CANYON CREEK STONECROP, PARADISE STONECROP
Sedum paradisum subsp. *paradisum* — 5 symmetrical petals forming a tube
Stonecrop family

After years of seeking this wildflower I finally connected with this interesting little stonecrop only to realize that it had always been right there under my nose in a place I am familiar with–the balds above the Sinks along Canyon Creek. It is small as far as stonecrops go–most of those I found were around 8" tall and blooming in open, sunny, and dry places. The creamy white flowers never fully open, as with other stonecrops, but retain a tight tube shape only flaring at the petal tips which surround a cluster of bright yellow anthers. The thick, reddish-green flowering stalk emerges from small, tight rosettes of gray-green fleshy leaves that are tinged with pink. Because of its size and scattered nature most backpackers would pass right by without noticing it. **Note:** I prefer the older name, paradise stonecrop, to the commonly accepted name. The location lends itself to the older appelation.

POPCORN FLOWER, CRYPTANTHA
Plagiobothrys species and *Cryptantha* species
Borage family

CALIFORNIA MISTMAIDEN
Romanzoffia californica
Waterleaf family

CREAM STONECROP
Sedum oregonense
Stonecrop family

CANYON CREEK STONECROP
Sedum obtusatum subsp. *paradisum*
Stonecrop family

CALIFORNIA VALERIAN
Valeriana californica
Valerian family

5 symmetrical petals foring a tube

This pretty little valerian is similar to its close relative *V. sitchensis* in most aspects except for size. This one remains fairly small (12" to 18") with a stem and leaf structure similar. The creamy white bunches of flowers appear identical but smaller with the same overall look sporting prominent stamens and anthers. It makes itself right at home, tucked away in the low growing stands of sage brush around East Boulder and Mill Creek Lakes in the Trinity Alps. I don't know why they favor hunkering down within the stands of sage but maybe it has something to do with protection from the persistent winds that funnel through this area from the Scott Valley in the north to the valley of the Trinity River in the south. I photographed these in mid-June and fought the wind at every release of the shutter!

MOUNTAIN HELIOTROPE, SITKA VALERIAN
Valeriana subsp. *sitchensis*
Valerian family

5 symmetrical petals forming a tube

This valerian is, literally, a show stopper. It is tall (2' - 4') and often towering over other plants in the near vicinity which really sets it apart. It is also distinguished by the large, beautiful clusters of snow-white flowers and pink tinged buds blooming at the tops of the flowering stalks. Stamens and creamy pink anthers are prominent. All of this makes it readily identifiable. These wildflowers prefer moist, shaded sites at the edges of meadows, near springs or other running water where it can bloom in abundance. The coarsely toothed leaves are cauline meaning that they grow mostly along the upper part of the stout stems. I photographed these at the edge of the meadow at Lower Albert Lake in the Russian Wilderness in September.

CALIFORNIA VALERIAN
Valeriana californica
Valerian family

SITKA VALERIAN
Valeriana sitchensis
Valerian family

CALIFORNIA HESPEROCHIRON
Hesperochiron californicus
Waterleaf family

5 symmetrical petals joined into a bell

This is a wildflower of open, wet to boggy meadows. It blooms as the snow melts at mid to high elevations, often in large colonies. A single, 1" in diameter, bell-shaped white or light pink flower blooms on a 2" to 3" stalk rising from a basal rosette of 6 to 8 leaves. The leaves are hairy, olive-green, and lay mostly flat to the ground. I shot the large photograph on Scott Mountain in mid-July and the smaller photograph in Morris Meadows in mid-June. They occur commonly throughout the area.

ALPINE SPICY WINTERGREEN
Gaultheria humifusa
Heath family

5 petals joined into a bell

"Expect the unexpected and you will never be disappointed." It was my Dad's mantra and really hit home when I knelt down to fill my cooking pot with water from a small lake in the Russians a few years ago. I caught a quick reflection of red and white in the water. I took a closer look, and then even closer, it turned out I had found a wildflower I'd never seen before. It frequents the shorelines of lakes and streams and is only a few inches tall. The flowers are pea sized or smaller, snow white, and fleshy with deep red sepals and dark orange pistils set against a green center. They generally hang downward from stout little stems. The leaves are oval in shape, deep green, tough, and leathery. I took these photographs while lying in the lake, with my camera on a small tripod which was also sitting in water. To get the photographs I had to shoot almost straight up into the small plant where it hung just above the water. I've visited this lake many times and filled my cook pot in the same spot over and over again which leads me to the conclusion that I definitely need to be more observant even when doing mundane chores. My Father would be proud.

SHORT SPURRED PLECTRIS, PINK PLECTRITIS
Plectritis congesta subsp. *brachystemon*
Valerian family

5 petals joined into a bell

To the casual observer there is not much to differentiate this little plectris from *P. macrocera* but if you get nose to nose with it, and maybe have a magnifying glass handy, you will notice a tiny spur extending outward below the lower lip of the lower petal which lies just outside the corolla tube. The corolla tube itself is behind the spur and descends down into the calyx. Other than the spur the two appear pretty much the same as far as structure goes. I photographed these in a meadowy area above Lewiston Lake in mid-May.

LONGHORN PLECTRITIS, WHITE PLECTRITIS
Plectritus macrocera
Valerian family

5 petals joined into a bell

This attractive little member of the Valerian family grows straight as a post from 4" to 14" tall. The opposite leaves are eliptical with short petioles and are widely spaced along the stems. The beautiful little tubular shaped white flowers flare away at the tips which are tinged with light pink. As in the other valerians, stamens are prominent. The colonies I've found prefer open, sunny sites and are common along certain stretches of the road to the Canyon Creek Trailhead in the Trinity Alps. It is common and widespread throughout the lower elevations of these mountains. I photographed these in early April.

CALIFORNIA HESPEROCHIRON
Hesperochiron californicus
Waterleaf family

ALPINE SPICY WINTERGREEN
Gaultheria humifusa
Heath family

SHORT SPURRED PLECTRIS, PINK PLECTRIS
Plectritis congesta subsp. *brachystemon*
Valerian family

LONGHORN PLECTRITIS, WHITE PLECTRITIS
Plectritis macrocera
Valerian family

WESTERN YARROW
Achillea millefolium
Sunflower family

Flowers in composite heads

This is a common and wide-spread wildflower of open, damp or dry meadows and slopes. It can be found in bloom throughout the entire season from May into late October in some areas. The fern-like leaves are pleasantly aromatic, pinnately divided, somewhat hairy, and grow smaller as they ascend the flowering stalks. The ray flowers are white and surround a small group of tiny yellowish disc flowers in a slightly flat-topped cluster. I photographed these along the Paynes Lake Trail in the Russians in mid-August.

PEARLY EVERLASTING
Anaphalis margaritacea
Sunflower family

Flowers in composite heads

Pearly everlasting comes by its name honestly. It blooms early and the flower heads persist well into the late fall season. It is a common trailside wildflower and is often the last bloom you will see before snow comes to the high country. I have no doubt that beneath the first snows of winter, you could still find flowers looking as fresh as if they bloomed in early July—it is the white involucral bracts of the heads that stay so pristine. The individual flowers resemble pussy toes with the same basic shape. The plants are fairly tall but loosely branched and the weight of the flower heads often bends the stalks to the ground. I photographed these along the Canyon Creek Trail near the Sinks in late September, but you will find it throughout this area alongside backcountry trails from the coasts to the mountains.

ALPINE PUSSYTOES
Antennaria media
Sunflower family

Flowers in composite heads

This little pussytoe is a common sight in the high country and easily recognized, even when not in bloom, by its small, low growing rosettes of fuzzy, pale gray-green leaves. It spreads across open sandy areas and often there is nothing else growing in the near vicinity. The fuzzy flowers that do, indeed look like cats toes and bloom atop a tall stalk, usually about 6" in height. I photographed these in the rocks above Big Blue Lake in the Russian Wilderness in September. Thousands of them populate open areas above Big Caribou Lake in the Trinity Alps.

SAWYER'S PUSSYTOES
Antennaria sawyeri
Sunflower family

Flowers in composite heads

As if there weren't enough cats in the high country here's one more—a brand spankin' new one! A few years ago I came across Pete Figura—biologist with the California Department of Fish and Wildlife—crawling around in the dirt near Stonewall Pass studying this little pussytoe. He indicated that it might be a new species and, as it turns out, he was correct. The plants are mat forming, about 7" tall with gray-green leaves and stems covered with fine whitish fuzz. Leaves alternate along the entire length of the flowering stalk and are lance to somewhat egg shaped. Individual plants are either male or female. In this photograph the flowers are beginning to turn brown with age but early on they are creamy white. I took this photograph near Stonewall Pass in early August.

WESTERN YARROW
Achillea millefolium
Sunflower family

PEARLY EVERLASTING
Anaphalis margaritacea
Sunflower family

ALPINE PUSSYTOES
Antennaria media
Sunflower family

SAWYER'S PUSSYTOES
Antennaria sawyeri
Sunflower family

GREENE'S BRICKELLBUSH
Brickellia greenei
Sunflower family
Flowers in composite heads

This is one of those wildflowers that you will probably smell long before you see the actual blooms. To my nose it's sharply obnoxious, but some people find it pleasant. Either way, you know immediately when it's in the proximity. It is a fairly common wildflower at lower elevations in the Trinity Alps and elsewhere. It grows in large colonies in rocky, open sites and, locally, is found mostly on serpentine soils. The 6" to 24" tall stems are covered with sticky hairs and topped with creamy-white rayless flower heads. The numerous style branches, two per flower, are long and give the flower heads a ragged look—overall though the flowers are pretty. The photograph of these flowers was taken along the lower reaches of the Bear Creek Trail in the Trinity Alps in August.

FREMONT'S CALYCADENIA, FREMONT'S WESTERN ROSINWEED, KLAMATH CALYCADENIA
Calycadenia fremontii
Sunflower family
Flowers in composite heads

This wildflower can reach to 40" in some places but 12" to 16" is more the norm. Stems are brown, strongly erect with alternate, linear shaped leaves which appear bunched along the stems. The leaves and stems are hairy. The slightly reflexed ray flowers are generally white (sometimes aging to pink) but can also be cream or yellow. The disc flowers are few and somewhat elevated—each with a single protruding purple-black anther tube. The leaves are somewhat sticky to the touch and smell slightly of varnish. I photographed these along the Rush Creek Road near Lewiston in September.

DUSTY MAIDENS, HOARY PINCUSHION, HOARY FALSE YARROW
Chaenactis douglasii
Sunflower family
flowers in composite heads

This little chaenactis grows in low, spreading mats with small, compact leaves that are fleshy and a dark green. The greenish-red flowering stems are mostly erect, fairly tall (around 7" or so), and leggy. The compact heads of magenta flowers gradually turn white as they open. Each flower has five recurved and sharply pointed petals surrouding a group of prominent stamens. These flowers were along the east side of the Parks Creek Road where the PCT crosses in mid-June.

SHASTA PINCUSHION, SHASTA CHAENACTIS
Chaenactis suffrutescens
Sunflower family
Flowers in composite heads

The plants in this fairly large population were growing in low, rounded mounds on a rocky, steep, exposed serpentine road-cut. The deeply cleft, fern-like leaves are silvery-green, somewhat fuzzy, and sticky, as are the flowering stalks and flower heads. The small, ½" to ¾" diameter, cream-colored flowers are discoid, meaning there are no ray flowers, so there are no obvious "petals" on the flower heads. Each flower head sits solitary on a 3" to 6" stalk well above the leafy mound. The round seedheads are no less beautiful—made up of hundreds of translucent white almost petaloid pappus scales atop the fruits. Julie Kierstead and Gary Nakamura, in their book "Selected Rare Plants of Northern California," list Shasta pincushion as rare and endemic to this area. It seems to favor areas of ultramafic (serpentine) geology. I photographed these at the North end of Trinity Lake along Trinity Mountain Road in late May.

GREENE'S BRICKELLBUSH
Brickellia greenei
Sunflower family

FREMONT'S CALYCADENIA
Calycadenia fremontii
Sunflower family

DUSTY MAIDENS, HOARY PINCUSHION
Chaenactis douglasii
Sunflower family

SHASTA PINCUSHION, SHASTA CHAENACTIS
Chaenactis suffrutescens
Sunflower family

PEREGRINE THISTLE
Cirsium cymosum
Sunflower family

Flowers in composite heads

This is one of the easier wildflowers to identify. It grows quite tall and sometimes reaching 5', with slender stems and leaves that clasp the stalks—all covered with white cobwebby hairs. What really gives it away are the loose bunches of creamy white flower heads that emerge from a rounded, spiny collection of gray-green sepals. It prefers semi-moist environments at mid to higher elevations, often in serpentine or other rocky areas. I took this photograph alongside the road to Kangaroo Lake on the Trinity Divide in July. Scott Mountain is another place to find them.

DWARF THISTLE, ELK THISTLE
Cirsium scariosum
Sunflower family

Flowers in composite heads

This particular thistle can pose an identification problem for most amateurs. Its habit depends largely on elevation and, to some degree, on the substrate it is found on. Around the Deadfall Lakes Basin and the trail to the summit of Mt. Eddy the plants are prostrate ground huggers while in other areas it can be fairly tall. The colors of the flowers are also variable—from grayish white to magenta. That being said, if you find yourself in its presence along the northern end of the Trinity Divide you can be fairly assured that this is what you are looking at. The deep green leaves are cleft all the way to the cream-colored mid-rib and are spiny—no fun to handle! I photographed this one on the open, dry serpentine slopes of Mt. Eddy in mid-July amidst a profusion of other wildflowers. I found them blooming all around the summit at a bit less than 9000'

CUT LEAVED DAISY, CUTLEAF DAISY
Erigeron compositus
Sunflower family

Flowers in composite heads

This pretty little daisy grows in small mounds on open, exposed talus slopes at higher elevations around the north end of the Trinity Divide, the Trinity Alps, the Russians, and Marble Mountain wilderness areas. The leaves are tiny, soft, somewhat fleshy, minutely pubescent, and tightly packed. A single flower head blooms atop a fleshy stem 4" to 6" tall. Ray flowers are white to light lavender surrounding the numerous golden yellow disc flowers. I found and photographed these along the trail above Upper Deadfall Lake and around the summit of Mt. Eddy, which is just a hair over 9000'. They were common throughout the area and in prime time for the bloom in the middle of July.

PEREGRINE THISTLE
Cirsium cymosum
Sunflower family

DWARF THISTLE, ELK THISTLE
Cirsium scariosum
Sunflower family

CUT LEAVED DAISY
Erigeron compositus
Sunflower family

OX-EYE DAISY
Leucanthemum vulgare
Sunflower family

Flowers in composite heads

This naturalized European immigrant is a common road-side flower found throughout Northern California—from the foothills of the Central Valley to the Coast. I've not found it inside the Trinity Alps, Russian or Marble Mountain Wilderness areas but I do see it along the roads to several of the trailheads. It is readily identifiable with its large white ray flowers and the bright yellow "eye" of disc flowers in the center, and its pungent odor. The stems can be up to 2′ tall. Its leaves are toothed and wider at the tip. Lower leaves are tapered to a long petiole. Leaves on the stem are mostly sessile. Look for it along Highway 3 in the Trinities and the road to the Duck Lakes Trailhead on the east side of the Russians.

LITTLE-LEAF SILVERBACK LUINA, LITTLE-LEAF LUINA
Luina hypoleuca
Sunflower family

Flowers in composite heads

In mid-May along Trinity Mountain Road/East Shore Drive (one becomes the other at the East Fork of the Trinity River) near the north end of Trinity Lake, I spotted this attractive plant growing in a crevice on a steep and rocky road cut or cliff face. At that time the leaves were fully formed into a beautiful silvery-green mound about 30″ across but the flowers were still just tiny white buds. On several subsequent trips into the same area I kept close watch as I wanted to catch it in full bloom. Exactly one month later, in mid-June I was rewarded with fully opened flowers. The leaves are almost egg shaped, green above and covered with white-wooly hairs below, mostly entire or slightly toothed and alternate along erect stems. Clusters of disc flowers emerge at the ends of the stems and are creamy white. Five yellow-orange anthers fused into a tube protrude well beyond each disc flower and at the end of each tube a 2-branched style emerges. There are no ray flowers.

WESTERN SWEET COLTSFOOT
Petasites frigidus var. *palmatus*
Sunflower family

Flowers in composite heads

Coltsfoot is fairly common throughout this region at lower to mid-elevations. It likes its feet wet and prefers shade. Look for it along streams and in springs and seeps in areas that are mostly shaded during the day. First thing in the spring, usually in early April, the thick flowering stalks emerge and can grow to 24″ or more. The large, palmately lobed leaves emerge after the stalks but before the large heads of white to lavender flowers begin to fully open. As with most wildflowers, the closer you get to these the more beautiful they appear. The individual creamy white tubular flowers have five petals that flare away at the tips and surround the styles that are sheathed in the deep purple anthers. I photographed these along Canyon Creek Road in mid-April.

OX-EYE DAISY
Leucanthemum vulgare
Sunflower family

LITTLE-LEAF SILVERBACK
Luina hypoleuca
Sunflower family

WESTERN SWEET COLTSFOOT
Petasites frigidus var. *palmatus*
Sunflower family

WOOD ANEMONE, COLUMBIA WINDFLOWER
Anemone deltoidea 5 symmetrical petals - ovary superior
Buttercup family

The large white "petals" of this anemone aren't really petals at all but sepals that surround the yellowish-green ovary which, in turn, is surrounded by large numbers of fine, yellowish star-like stamens. As with all anemones featured in this guide there are no true petals. This is a delicate plant that doesn't grow much over 8" tall with a whorl of three, wavy edged leaves below the flowers and just above the mid-point of the stem. Its preferred habitat is shaded, moist areas and is one of those wildflowers that make the hike into the Duck Lakes Basin so delightful. I photographed these along the trail to Big Duck Lake in the Russian Wilderness but they are also common along the trail to Granite Lake in the Trinity Alps.

WESTERN PASQUEFLOWER, EINSTEIN HEADS
Anemone occidentalis Many petals
Buttercup family

Because, late in the season, the flowers of this anemone give way to big rounded, shaggy masses of plumose-tailed achenes (seeds to us amateurs) they've come to be referred to as "Einstein heads." The resemblance between the scientist and the seed head is pretty amazing. The flowers push their way through receding snowbanks and are some of the earliest bloomers in the high country. Like the other anemones, the white petals are not petals at all, but petal-like sepals about 1 1/4" long, surrounding masses of yellow-orange stamens and greenish pistils. A good place to see them is at the upper end of Long Canyon near the chute leading to Lake Anna where they mass in fields of white. I photographed the flowers in Long Canyon in the Trinity Alps and the seedheads on Mt. Ashland in southern Oregon.

DRUMMOND'S ANEMONE
Anemone drummondii var. *drummondii* Many petals
Buttercup family

This is THE flower of open alpine and subalpine slopes throughout the entire area covered by this guide. It is one of the earliest bloomers in the high country, often peeking out from beneath snowbanks as they recede in the spring. Entire hillsides can be covered with them. Each plant has several stems up to 12" tall. The leaves are finely dissected and hairy. Each stem sports a solitary white flower. The white "petals" are actually sepals that surround a mass of tiny yellow stamens clustered around multiple greenish pistils in the flower's center. Drummond's anemone is easily recognized by the soft blue tinge appearing on the underside of the sepals shown in the small photograph taken above East Boulder Lake in June. The large photograph was taken on Parker Divide in the Trinity Alps in late May.

OREGON ANEMONE, BLUE WINDFLOWER
Anemone oregana var. *oregana* 5 symmetrical petals - ovary superior
Buttercup family

Most manuals more or less describe this anemone as blue. I've looked high and low in these mountains and have yet to find a blue one. They have been consistently white. So rather than leaving it in the blue section and as we did in the 1st edition I've now moved it to here its new more relevant location. It is characterized by a single basal leaf that is divided into three and then divided into three again. Just below the flowers you will find three-parted bracrts that look like leaves. The "petals" you see are actually sepals and are about 1 1/4" in diameter. Stamens usually number more than 35. I photographed these on Scott Mountain in the mid-May. I've also photographed them along the road to the North Fork of Swift Creek Trailhead in late April. If you find blue versions it would be nice to know when and where.

WOOD ANEMONE, COLUMBIA WINDFLOWER
Anemone deltoidea
Buttercup family

WESTERN PASQUEFLOWER
Anemone occidentalis
Buttercup family

DRUMMOND'S ANEMONE
Anemone drummondii var. *drummondii*
Buttercup family

BLUE WINDFLOWER, OREGON ANEMONE
Anemone oregana var. *oregana*
Buttercup family

MARSH MARIGOLD
Caltha leptosepala
Buttercup family

Many petals

This absolutely gorgeous wildflower is one of the earliest species to emerge in the high country, often peeking from beneath snow banks as they recede from May to early June. The white "petals" are actually sepals—each about 5/8″ long. The flowers bloom at the ends of stout stalks and are topped with numerous yellow stamens. The dark green basal leaves are oblong to round and are somewhat folded along the central axis. It is common to find them blooming alongside water plantain buttercups (page 238). Huge colonies of them grow in wet, marshy places like Upper Union Meadows just below Parker Divide. Most people are surprised to learn that marsh marigolds have no petals. I photographed these in the meadows between West Boulder and Telephone Lakes in July.

NEVADA LEWISIA
Lewisia nevadensis
Miner's lettuce family

Many petals

Nevada lewisia can be found at mid-elevations in damp meadows throughout the Trinity Alps, the Marble Mountain, and the Trinity Divide. I have not found it in the Russians. The plants and flowers are tiny, only a couple of inches tall, and hide in newly emerging meadow grass, making them sometimes difficult to spot. The small, deep green leaves are linear in shape and fleshy. Flowers are white with olive green throats and streaks of green running from the throat to the petal tips. Unless you are attentive as you are walking you are not likely to find them. I photographed these in Union Meadows in late May and Cement Bluff Lake in early July. **Note:** Flowers close up in the evening as daylight fades, fully opening again as morning breaks.

THREE-LEAVED LEWISIA
Lewisia triphylla
Miner's lettuce family

Many petals

This tiny lewisia is another wildflower that hides in plain sight. The snow-white flowers are typically less than 1/4″ across and the entire plant will hardly ever grow beyond a couple of inches in height. Stamens are somewhat prominent and the anthers bright pink to magenta. The narrow leaves are fleshy and curve up and away at the base of the flowering stem. A few years ago, walking to Little Granite Lake from Doe Lake in the Trinities, I found that I had overlooked large populations of them while traversing the storm damaged trail. I shot these photographs in the wet meadows surrounding East Boulder Lake in the Trinity Alps in mid-June. They are wide-spread throughout the Trinities and Russians.

MARSH MARIGOLD
Caltha leptosepala
Buttercup family

NEVADA LEWISIA
Lewisia nevadensis
Miner's lettuce family

THREE-LEAVED LEWISIA
Lewisia triphylla
Miner's lettuce family

BITTERROOT
Lewisia rediviva Many petals
Miner's lettuce family

All lewisias are gorgeous, but for sheer size and beauty this one takes the proverbial cake. The large and showy flowers crown what, at first glance, seems to be relatively "barren" and open serpentine ground like a carpet in late spring and early summer. The translucent snow-white flowers emerge from a fleshy taproot and are tinged from beneath by pink to magenta sepals. Numerous white filaments topped with bright pink anthers comprise an array of stamens. I photographed these on the slopes of Scott Mountain in a year when the bloom was quite extensive. There were so many of them it was hard to walk in some places without stepping on them. If you are into such things, it is well worth the drive and relatively short walk just to see these and the myriad other species that grace this wildflower hot spot.

Historical note: I love this traditional Shoshone and Flathead Indian story about bitterroot: "One day an elderly woman couldn't find food for her family and started crying. The sun sent a red bird to comfort her. The bird told her that every place her bitter tears fell, a root, nourished by her love, would grow. The root and flower would be colored by the woman's white hair. The taste would be flavored by her bitter tears and the flower would be tinted by the bird's red feathers and she would always have food for her tribe."

Bitterroot, or "spetlum" as the Flathead called it, is an important part of their diet, but considered more a luxury or delicacy as opposed to a staple. The Flathead and neighboring tribes timed their spring migrations with the blooming of the bitterroot. May is known as "bitterroot month." The Shoshone believe the small red core found in the upper taproot has special powers, notably being able to stop a bear attack. The roots also has several medicinal uses.

Western science named the plant in honor of Meriwether Lewis, who collected it during the Lewis and Clark Expedition and brought it back for study. Bitterroot is the state flower of Montana, where Lewis made his discovery.

COTTON GRASS
Calliscirpus criniger no obvious petals
Sedge family

This easily recognized member of the sedge family frequents wet meadows, seeps and fens throughout the high country. The flowering heads are showy and quite distinctive. They can be a couple inches in diameter and made up of hundreds of stamens topped with golden yellow anthers. The flowers bloom at the tops of stems that are usually around 10" to 14" tall. The large photograph was taken in a wet meadow alongside Highway 3 where it passes over the summit of Scott Mountain and the smaller one in the meadow below Deadfall Lakes on the Trinity Divide. I've also found it around Taylor Lake in the Russians and in Van Matre Meadows and Gibson Meadows, both in the Trinity Alps .

BITTERROOT
Lewisia rediviva
Miner's lettuce family

Sepentine and widlflowers - Scott Mountain

COTTON GRASS
Calliscirpus criniger
Sedge family

WESTERN SERVICEBERRY

Amelanchier alnifolia — Alternate leaves - 5 symmetrical petals ovary superior

Rose family

This shrub is common and found throughout these mountains. It can be found singly but often forms dense thickets on semi-dry rocky hillsides and flats. The flowers are quite attractive, snow white, ruffled at the margins, and blooming in tight clusters all along the branches. In the fall they yield a dry, blue fruit that resembles a small apple in shape. The leaves are oval and toothed along the upper half. The berry clusters are often covered with orange rust. I photographed these along the trail to Lilypad Lake in the Trinity Alps in July.

MOUNTAIN WHITETHORN

Ceanothus cordulatus — 5 symetrical petals - ovary inferior

Buckthorn family

This ceanothus is a stout, thorny shrub that grows in dense thickets in dry, exposed places. Cross country travel through thickets of mountain whitethorn should be avoided at all cost as it can be a painful and bloody experience. They are also good places to pick up oodles of deer ticks, which love this shrub. The leaves are mostly oval and, like other ceanothus species, have three prominent veins. The small, white flowers bloom in small, dense clusters up to 2″ across. The flowers of this particular ceanothus impart an almost sickly sweet smell that really gets my allergies going when I run trails where it blooms in the spring. I photographed these along the road to the Stoney Ridge Trailhead.

BUCKBRUSH, WEDGELEAF CEANOTHUS

Ceanothus cuneatus — 5 symmetrical petals - ovary inferior

Buckthorn family

This is a common ceanothus throughout the foothills of the area covered by this guide, and the earliest ceanothus to bloom in the spring. The oppositely arranged branches are stiff, light gray, and almost white. The gray-green leaves are leathery, evergreen, about 3/4″ long and mostly obovate. Small white flowers bloom in loose clusters along the stems and, like most of the shrubs in the ceanothus family, have an almost overpoweringly sweet smell that triggers a lot of allergic reactions in people. It gets its common name from the fact that it (like others in the ceanothus family) is a favorite foraging shrub for deer especially in winter. I photographed these on Bowerman Ridge above Trinity Lake.

SNOWBRUSH, TOBACCO BRUSH, VARNISH LEAF

Ceanothus velutinus — 5 symetrical petals - ovary inferior

Buckthorn family

Snowbrush is a high country favorite of mine. The large leaves, when warmed by the sun, emit a wonderful aromatic and spicy smell. My first experience with it was on a day hike into Caribou Lake in the 1950s where the open, south facing slopes of the old trail were covered with it. Dad would occasionally pick a few leaves and put some beneath his pack straps where the constant rubbing accentuated the smell. Take some time on your next outing and breathe in some of this delicious aroma! As the sun warms the slopes where it grows, the "sap" they exude clings like varnish to the tops of the leaves, giving it the other common name, varnish leaf. The creamy white flowering bunches look fluffy from a distance and can almost completely bury the branches and leaves. I photographed these near the top of the ridge on the Old Caribou Trail in the Trinity Alps in August.

WESTERN SERVICEBERRY
Amelanchier alnifolia
Rose family

MOUNTAIN WHITETHORN
Ceanothus cordulatus
Buckthorn family

BUCKBRUSH, WEDGELEAF, CEANOTHUS
Ceanothus cuneatus
Buckthorn family

SNOWBRUSH, TOBACCO BRUSH, VARNISH LEAF
Ceanothus velutinus
Buckthorn family

A NOTE ABOUT COFFEEBERRIES: Several species and subspecies of coffeeberry are found throughout this region including the two illustrated here. It is a common shrub or small tree which can grow to about 15′ in height depending on elevation, soils, and the surrounding micro-climate. In the previous edition of this guide I noted that it prefers dry areas but I've found that not to be exactly true. It also grows next to springs and along the banks of streams but it does not necessarily have its feet in water. Differences between the species and subspecies will probably not be apparent to the casual on-looker because of their similar habit and overall appearance. Suffice it to say—if you find shrubs or small trees that are fairly consistent with the photographs and descriptions below you have found yourself a coffeeberry!

CALIFORNIA COFFEEBERRY
Frangula californica subsp. *occidentalis* — 5 symetrical petals - ovary inferior
Buckthorn family

Leaves of this species are elliptical, 2″ to 3″ long, with prominent pinnate veins and smooth leaf margins—deep green above and yellowish beneath. The greenish-white flowers are small, star shaped, and grow in inconspicuous clusters. The fleshy fruit is small, about 3/8″ in diameter, and turns a deep purple black in late summer. It has a surprisingly pleasant coffee flavor and was used by early settlers in lieu of coffee at times. I photographed the flowers along the Trinity River in mid-June and the fruit in the same location in early September.

SIERRA COFFEEBERRY, RED BUCKTHORN
Frangula rubra — *5 symmetrical petals - ovary inferior*
buckthorn family

This shrubby small tree is, in most ways, similar to California coffeeberry, the most noticeable difference being its striking red to reddish green stems. It seems to do well in drier, exposed areas where it gets a healthy dose of sun—perfectly suited to this area's climate. Leaves are elliptical, dark green, prominently veined, and slightly toothed on the wavy margins. They range in height from 8′ to about 15′ feet. The fleshy fruit occurs in bunches and is deep purple-black. I took these photographs just above Lewiston Lake in mid-May.

CALIFORNIA COFFEEBERRY
Frangula californica subsp. *occidentalis*
Buckthorn family

SIERRA COFFEEBERRY, RED BUCKTHORN
Frangula rubra
Buckthorn family

California coffeeberry growing along old Highway 3 just above the Trinity River. This shrub is loaded with ripening berries.

TRAILING GOOSEBERRY, GROUND GOOSEBERRY

Ribes binominatum
Gooseberry family

5 symetrical petals - ovary inferior

This is a small, ground hugging gooseberry that forms low tangled mats that spill over rocks and downed woody debris in fairly dry and open meadows. The woody stems are spiny (spiny species are called gooseberries while the unarmed species are called currants) and no fun to wade through unless you're wearing boots. The dark green leaves are small—2" to 3" across— lobed, toothed and slightly fuzzy. The small white to greenish-white flowers are sticky and bloom in small clusters which, later on in the season, produce spiny (but sweet) fruits. The fruit has a nice flavor but warning—gloves are absolutely required if, perchance, you would like to sample one! I photographed this one in Union Meadows in early June where they grow in abundance.

STICKY CURRANT

Ribes viscosissimum
Gooseberry family

Alternate leaves - 5 symmetrical petals - ovary inferior

This is a fairly common, loosely branched shrub, usually about 4′ to 5′ in height, growing at the edges of meadows and other mixed conifer openings at mid-elevations. It is common and widespread throughout these mountains. The leaves are stiff, dark green, deeply veined, and shaped like maple leaves. Flowers are white, lightly tinged with pink, somewhat tubular in shape, with the sepals flaring at the tips. I photographed the flowers at the outlet of Mill Creek lake in mid-June and the inset of the fruit in Grizzly Meadows in September both in the Trinity Alps. **Note**: Julie Knorr insists that I mention the fragrance because, as she says, "it's my favorite part of the plant!"

Union Meadows in the Trinity Alps is largely ignored by backpackers on their way to the lakes on beyond. It is, however, attractive to horsepackers because of the abundant grass. I largely avoid the lakes because of the crowds they attract instead choosing to make them a dayhike. These meadows stretch from Union Creek below Union Lake all the way to the Parker Divide and beyond—providing habitat for a myriad of wildflower species and lots of wildlife. It's a great place for late spring and early summer thunderstorms also, which makes for outstanding photography.

TRAILING GOOSEBERRY
GROUND GOOSEBERRY
Ribes binominatum
Gooseberry family

STICKY CURRANT
Ribes viscosissimum
Gooseberry family

Union Meadows looking up to Foster Lake

BLUE ELDERBERRY

Sambucus nigra subsp. *caerulea* — 5 symmetrical petals - ovary inferior

Muskroot family

This large fruiting shrub is found throughout the area covered by this book from the foothills to mid-elevation meadows and hillsides. The bushes can reach heights of 16′ or more depending on location. The creamy white flower heads are large, compact and have a pleasant smell. The fruit turns a dark blue by late summer and is often covered with a white, waxy bloom. The fruit of the blue elderberry is seedy but excellent for eating and can be included in pancakes, baked in pies, made into syrup and jams, and even used to make wine—but first you have to beat the critters to it. It is a favorite snack for bears, birds, chipmunks, and pine squirrels. I shot these photographs in Bear Basin, the small photograph along Canyon Creek, both in the Trinity Alps. I found the fruit along the road to the Deacon Lee Trailhead in the Russian Wilderness.

RED ELDERBERRY

Sambucus racemosa var. *racemosa* — 5 symmetrical petals - ovary inferior

Muskroot family

This shrubby elderberry is commonly found at mid to higher elevations throughout the high country and is more often associated with damp areas like the shorelines of lakes and streams or near the edges of marshy meadows. The snow white flowers bloom in a small and compact bunch at the ends of the flowering stems. Late in the summer season, bright red berries ripen in large clusters but, unlike those of their cousin the blue elderberry, are NOT edible. Even though they look absolutely delicious hanging there from those beautiful purple stems avoid them except to take pictures. I photographed the flowers on the shoreline of Buck Lake in August and the berries on the shoreline of Sugarpine Lake in September. I've also found them in full bloom at Sugarpine in early October. Both locations are in the Trinity Alps.

The ABCD Lakes in the Marble Mountain Wilderness: Also known individually as Aspen, Buckthorn, Chinquapin and Dogwood Lakes, sit deep in a bowl surrounded by peaks and ridges with beautiful views of the scenery to the Northwest. A stiff climb from the lower lake takes you down to Deep and Wolverine Lakes and the gorgeous wildflower filled meadows that surround them. I found plenty of opportunities for photography in this stunning setting.

BLUE ELDERBERRY
Sambucus nigra subsp. *caerulea*
Muskroot family

RED ELDERBERRY
Sambucus racemosa var. *racemosa*
Muskroot family

The ABCD Lakes in the Marble Mountain Wilderness

SHASTA SNOW WREATH
Neviusia cliftonii
Rose family

No obvious petals

Snow wreath occupies soils derived from limestone and metasedimentary rocks and can be found growing in cool, well shaded stream drainages in the hills surrounding Shasta Lake. This part of the eastern Klamath, called the McCloud Range, is the only place on Earth this species is found. It's a fairly tall shrub, sometimes reaching 7', and gregarious, often growing in large thickets. Its branches are thin, almost spindly and covered with reddish-brown, shreddy bark. In her book Selected Rare Plants of Northern California (co-authored with Gary Nakamura) Julie Kierstead describes the flowers as "small white powder puffs of more than 50 stamens and yellow anthers with toothed sepals widest near the tips." There are occasionally one or two petals but they are inconspicuous and fall off soon after the flower opens. The "powder puffs" are approximately ½" in diameter. The small bright green leaves are toothed and pinnately veined. I shot this photograph along a popular trail near Shasta Lake in early April.

WILD MOCK ORANGE
Philadelphus lewisii
Hydrangea family

4 petals

This is a common shrub, up to 8' tall, loosely branched and covered with small clusters of gorgeous white flowers. The flowers generally have four petals but, occasionally five, with prominent, creamy yellow anthers that give the blossoms an almost star-like quality. Older branches are gray in color while the younger ones, those nearest the flowers, are generally reddish green. I've found it along streams and in other places close to water but also in exposed dry sites—it seems that it's not too picky about where it plants itself. A good place to see it is along Canyon Creek Road on the way to the trailhead (where I shot the large photograph) and then along Canyon Creek Trail to a point about half a mile below Middle Falls. When it's in full bloom the flowers give off a delightful fragrance reminiscent of orange blossoms, hence the name. It is common and widespread throughout these mountains.

SHASTA SNOW WREATH
Neviusia cliftonii
Rose family

WILD MOCK ORANGE
Philadelphus lewisii
Hydrangea family

Dry ridgetop meadows above Twin and Taylor Lakes in the Russians

KLAMATH MANZANITA

Arctostaphylos klamathensis 5 petals joined into a bell

Heath family

Within the range of this guide, this interesting little manzanita is found mainly on serpentine soils and, because it is not drought resistant like other manzanitas, always on north facing or deeply shaded slopes between 5,700′ and 6,500′. It is found with greenleaf and pine-mat manzanitas but, unlike these two with their bright green leaves, this one has grayish-blue-green leaves. Another distinguishing feature is its size which is intermediate between greenleaf and pine-mat—being between 1′ and 2′ tall. The other readily identifiable feature are glandular hairs that cover older twigs. New growth twigs are likewise hairy but each tipped with a sticky blob. I have to admit, I had a devil of a time finding this one so I could photograph it but once I took to heart the color of the leaves and size of the shrub it jumped right out at me. This species is only found within the range of this guide. I took these photographs on Scott Mountain in late April.

PINE-MAT MANZANITA

Arctostaphylos nevadensis 5 petals joined into a bell

Heath family

Though short in stature, this manzanita can cover a lot of ground. On open ridge tops at higher elevations it rarely gets taller than your knees. It favors sunny, dry hillsides, ridgetops, and the understory of conifer forests. The leaves are small, somewhat oval, finely veined, tough, and leathery. The reddish branches are likewise tough and resilient. The small white, urn-shaped flowers often have a pinkish tinge and occur in small clusters at the ends of branches. I shot the large photograph above the shoreline of Doe Lake in early August and the small one on Scott Mountain in late April.

GREENLEAF MANZANITA

Arctostaphylos patula 5 petals joined into a bell

Heath family

This manzanita is common throughout the high country—often forming dense patches that are almost impossible for a cross country hiker to get through without tearing the heck out of legs and clothing alike. It prefers dry, open, and sunny sites where it grows to a height of 5′. It is an irregularly branched shrub with deep red bark that peels away to reveal new bark beneath that is bright green to yellow. Leaves are deep green, tough, oval, and finely veined. The flowers resemble tiny, upside down white to deep pink urns. Mealy reddish-brown fruits ripen late in the summer and are favored by bears. I photographed the white to pale pink flowers along the Union Creek Trail in the Trinity Alps in early June and the pink flowers on Scott Mountain also in June.

WHITELEAF MANZANITA

Arctostaphylos viscida 5 symmetrical petals ovary inferior

Heath family

This might be the earliest bloomer on the block, at least in the shrub department. At lower elevations look for the bunches of white to pale pink flowers to open anytime after the middle of December. The leaves of whiteleaf manzanita are tough and leathery, darker green on the top side and a paler grayish-green beneath—which is how they got their name. As the shrub grows, its deep reddish-brown bark splits and reveals the new pale orange to yellowish-green bark beneath (see inset). The old bark curls up and eventually falls to the ground. The photograph of the flowers was taken above Lewiston Lake in late December.

KLAMATH MANZANITA
Arctostaphylos klamathensis
Heath family

PINE-MAT MANZANITA
Arctostaphylos nevadensis
Heath family

GREENLEAF MANZANITA
Arctostaphylos patula
Heath family

WHITELEAF MANZANITA
Arctostaphylos viscida
Heath family

LOVE HEATHER, WHITE HEATHER
Cassiope mertensiana 5 petals joined into a bell
Heath family

When this beautiful shrub blooms, it not only brightens up entire high elevation hillsides with its masses of dainty white, bell shaped flowers but it also brightens the mood. It typically is small, usually no more than1′ to 3′ tall, and thickly branched. Short, scale-like leaves encircle and hug each upright branch. It favors rocky outcrops and slopes, often gracing cliff faces. The nodding flowers, capped with bright red sepals, grow from the axils of the leaves and are approximately 1/4″ long. These photographs were taken at Big Duck Lake in the Russians but I've photographed them on the slopes above Salmon Lake and Upper Caribou Lake in the Trinity Alps as well.

SIERRA LAUREL
Leucothoe davisiae 5 petals joined into a bell
Heath family

This shrub grows throughout the high country in dense, waist high thickets along the shorelines of lakes, the edges of ponds, the banks of streams, and marshy areas. Thick clusters of small, pendulous, white, urn shaped flowers bloom on tall spikes from mid-summer well into the fall season. I have found them blooming as late as mid-October at Little Bear Lake in the Trinity Alps. The evergreen, oblong leaves are crowded along strongly upright and very flexible stems. I shot the large phtotograph along the shoreline of Big Boulder Lake and the close-up image along the stream below Doe Lake, both in the Trinity Alps.

CALIFORNIA SNOWDROP BUSH, SNOWDROP BUSH, STORAX
Styrax redivivus 5 petals joined into a bell
Storax family

This low elevation shrub has some of the prettiest flowers anywhere. Personally, I rank them right up there with the flowers of mock orange and Pacific dogwood for their graceful beauty. It is common in open, sunny areas or in partially shaded forest slopes. The snow-white flowers hang in pendulous bunches, the petals fused at the base and flaring open at the tips to expose bundles of white stamens crowned with elongated, creamy-orange anthers. Leaves are round, leathery, dark green with yellow-green veins, and wavy at the margins. In most instances it is about 4′ to 12′ in height. I photographed these on Bowerman Ridge in mid-June but in late spring there is quite a display along Highway 299W between Redding and French Gulch. **Note:** This shrub has an amazing variety of common names other than those listed above, among them—California styrax, bitternut, drug snowbell and chaparral snowbell. Take your pick!

LOVE HEATHER, WHITE HEATHER
Cassiope mertensiana
Heath family

SIERRA LAUREL
Leucothoe davisiae
Heath family

CALIFORNIA SNOWDROP BUSH
Styrax redivivus
Styrax family

COASTAL SILKTASSEL, WAVYLEAFED SILKTASSEL, QUININE BUSH
Garrya eliptica
Silktassel family

5 petals joined into a bell

This silktassel is common along the North Coast, in the coastal mountains, and less so farther inland. Like *G. fremontii* this one is a shrub that can also appear as a small tree. The dark green leaves are elliptical and strongly wavy—sometimes rolled under. The "tassels" are much longer than those of *G. fremontii*, sometimes close to 12" long with muted white with soft purple accents. The flowers are soft and fuzzy to the touch. The fruit appears the same. I shot these photographs in Patrick's Point State Park in late January.

FREMONT SILKTASSEL
Garrya fremontii
Silktassel family

5 petals joined into a bell

From a distance a silktassel bush looks a lot like a species of manzanita and most people pass it by as such. However, while its overall appearance might be similar a closer inspection reveals some obvious differences. For one, silktassels are not as stout as manzanitas meaning that wading through a patch won't leave you a bloody mess. For another, silktassel leaves are oppositely arranged along the branches while those of manzanita are alternate. What really gives them away are the flowers which hang in long, pendulous catkins from the ends of the branches (Silktassels are dioecious, meaning that any individual plant has all female flowers or all male flowers). Typically, the flowers are white and somewhat pubescent but It is interesting to note that shrubs standing shoulder to shoulder often have flowers that are different colors. Some, as illustrated, appear yellowish while right next door they appear almost purple. This difference is all in the color of the sepals. I've seen bunches of flowers at least 10" long if not longer, especially with the coastal specimens. In the fall they produce a tough, dark purple fruit much favored by bears. If you find yourself in the middle of a patch of ripening silktassel berries be on the lookout for the furry foragers with their purple lips. I shot both of these photographs within a few feet of each other on Weaver Bally in the Trinity Alps In late February and the photograph of the ripe fruit along the trail to Alpine Lake in September. If you make the hike up to the old Billy's Peak Lookout site above Coffee Creek in the fall when the silktassel is ripe you might be rewarded with herds of grazing bears. Years ago I counted at least a dozen of them rummaging around in the bushes along the ridge.

COASTAL SILKTASSEL, WAVYLEAF SILKTASSEL
Garrya eliptica
Silktassel family

FREMONT SILKTASSEL
Garrya fremontii
Silktassel family

THIN-LEAF HUCKLEBERRY, BLACK HUCKLEBERRY
Vaccinium membranaceum 5 petals joined into a bell
Heath family

I remember a warm October afternoon backpacking into Little Bear Lake—the air filled with the intoxicating smell of ripe huckleberries. Every shrub was loaded with these little blue treasures and the taste was every bit as delicious as the smell. There was a bumper crop of huckleberries that year with enough to go around for chipmunks, birds, bears, and humans alike. The shrubs can grow to 3′ or more and cover large areas. I photographed the slightly translucent flowers at Big Duck Lake in the Russians and the fruit at Little Bear Lake in the Trinity Alps. In the fall they put on a beautiful display of brilliant red foliage in places like the East Weaver Lake.

A note about huckleberries and billberries: Numerous species of huckleberry are found throughout this area and distinguishing one from the other is difficult, even for experts. Suffice it to say that you will likely run into any or all of them at different times and places as you hike these mountains. The two species illustrated in this guide (*V. cespitosum*, page 242) are pretty typical and maybe the most common. The fruit of all varieties is excellent eating and who really cares about technical things like names when there is the possibility of huckleberry pancakes on a chilly October morning in the high country?

COMMON BUTTONBUSH, BUTTONWILLOW
Cephalanthus occidentalis 5 symmetrical petals forming a tube
Madder family

If you are one of those people who enjoy canoeing or kayaking you are probably familiar with the striking flowers of this shrub found along the shorelines of low elevation lakes in this area. The ball-shaped floral display is made up of hundreds of tiny creamy white tube-shaped flowers out of which radiate long, slender pistils giving the over-all effect of an exploding star. Of course, you don't have to be on the water to enjoy this shrub because it frequents the banks of the Trinity River as well as other small streams. I've noticed that it is a favored landing spot for various dragonflies and butterflies find it particularly attractive as a food source. Some shrubs can reach heights of 10′ or so with deep green, egg-shaped leaves which are sharply pointed at the tips. I photographed these on Lewiston Lake in mid-July - from my canoe.

LABRADOR TEA, TRAPPER'S TEA
Rhododendron columbianum 5 symmetrical petals forming a tube
Heath family

Labrador tea is a waist high, stiffly branched shrub that decorates the shorelines of lakes and edges of streams throughout the high country—often in large and dense populations. The evergreen leaves are oblong, leathery, and crowded in tufts at the ends of the branches. White, five-petaled flowers bloom in rounded, compact clusters with conspicuous stamens. In protected pockets it can continue to bloom well into October. I photographed these around the shoreline of Big Duck Lake in the Russians in mid-July. **Note:** local Yurok Indians boil the leaves of Labrador tea to make a medicinal herbal tea. German brewers used a variety of it while brewing their beer which they believed made it more intoxicating (eventually it was forbidden by law because this practice supposedly led to increased aggression amongst imbibers).

THIN-LEAF HUCKLEBERRY
BLACK HUCKLEBERRY
Vaccinium membranaceum
Heath family

COMMON BUTTON BUSH
Cephalanthus occidentalis
Madder family

LABRADOR TEA
TRAPPER'S TEA
Rhododendron columbianum
Heath family

WESTERN AZALEA
Rhododendron occidentale
Heath family

5 symmetrical petals forming a tube

Every wildflower is beautiful, there can be no doubt, but this one takes the prize. Azalea flowers are large, showy and their fragrance is outrageously delightful. Walking through a patch of azalea along a high-country trail on a warm afternoon is nothing short of intoxicating and just about as close to heaven as you can get. In the Klamath Ranges, azaleas have an affinity with water and serpentine–where they prefer stream banks, the shorelines of ponds, lakes, and marshy areas. They can also be found growing in highly mineralized sites at low elevations especially around Keswick and Shasta Lakes–often where copper mining was prevalent years ago. You will find them blooming along the shorelines of Lewiston and Whiskeytown Lakes too. In the fall, leaves turn a beautiful gold to brilliant red. Goldenrod spiders (*Misumena vatia*) love the tubular shape of the flower because it provides a safe hiding place. They are aggressive little spiders that defend their flowery homes with all the ferocity of a miniature charging bear. The one pictured in the inset dashed out from his hidey-hole and attacked my camera lens every time I got to within a couple of inches of the flower. The large photograph was taken along the Union Lake Trail in early June and the crab spider along the North Fork of Coffee Creek in July, both in the Trinity Alps.

BITTER CHERRY
Prunus emarginata
Rose family

5 symetrical petals ovary superior

Bitter cherry grows throughout this area, often in impenetrable thickets on rocky, well drained hillsides. The branches are reddish. The leaves are oval, finely toothed, and about 1″ to 2″ long. The small, white flowers are grouped in little round-topped bunches at the ends of short branchlets. The fruit is a bright, almost translucent red, each about ¼″ in diameter and much favored by birds. If you want a memorable experience just pop a handful into your mouth and chew 'em up! They are mouth puckering bitter and I guarantee you won't do it again. The photograph of the flowers was taken in Union Meadows and the fruit on the slopes above Stoddard Lake, both in the Trinity Alps.

WESTERN CHOKECHERRY
Prunus virginiana var. *demissa*
Rose family

5 symmetrical petals - ovary superior

When I was a kid growing up in Lewiston I spent many late summer afternoon picking chokecherries for my mom and grandmother, who made big batches of jelly and syrup from the cherries. Beware though–they come by their common name honestly–eaten right off the tree they are decidedly mouth puckering! They can grow in thickets to 30′ tall in places with long racemes of white flowers decorated with crowns of deep yellow stamens. In late summer long bunches of deep purple cherries develop and, if you don't get to them first, are scarfed up by birds, squirrels, chipmunks, and bears. Look for them along streams where they can get their feet into water or on damp hillsides. I photographed the flowers in early June along Deadwood Creek, and the fruit at Lowden Ranch in early September both near Lewiston.

WESTERN AZALEA
Rhododendron occidentale
Heath family

BITTER CHERRY
Prunus emarginata
Rose family

WESTERN CHOKECHERRY
Prunus virginiana var. *demissa*
Rose family

DOG ROSE
Rosa canina
Rose family

5 symmetrical petals - ovary superior

Though not common throughout this area, there are several populations of this naturalized rose (originally native to Europe, northwest Africa, and western Asia) along the Trinity River. In places near Junction City it can be found climbing through brush and small trees. The showy snow-white flowers are structured like the other roses found in these mountains, having five petals with a concentration of yellow to deep orange anthers in the center and multiple tiny pistils enclosed in a hollow hip. They are usually 2½" to 3" in diameter and fragrant. I took this photograph along the river upstream from Junction City in early July. **Note**: Christina Wernham shot this gorgeous photograph of a red spotted crab spider (also known as "goldenrod spider" or "flower spider") on a dog rose near Dunsmuir along the Sacramento River.

THIMBLEBERRY
Rubus parviflorus
Rose family

5 symetrical petals ovary superior

One of my great joys when backpacking in the high country is finding ripe thimbleberries along the trail. If you are lucky enough to get to them before the birds and chipmunks they can bring any well planned hike to a complete and sudden halt. When ripe, the berries lift off like little reddish-purple beanie caps and the taste is outstanding. They are easily recognized with their large dark green, palmately divided leaves, and large, white, wrinkly-petaled flowers. They tend to grow in large colonies in shaded areas where there is abundant moisture. I shot these blooms near Enni Camp on the trail to Papoose Lake in August and the fruit along the Deacon Lee Trail in the Russians in September.

WESTERN POISON-OAK
Toxicodendron diversilobum
Sumac family

5 symetrical petals ovary superior

Ah... what can I say about poison-oak? If you don't know what it looks like by now, you are in for an itchy good time if you happen to wade through a patch of it. Remember the old maxim: "Leaves of three - let it be!" It is usually shrubby but can vine up surrounding trees to a height exceeding 40′. In early summer the deep green, oily leaves often turn a beautiful deep red which makes it a very attractive plant (I know a woman who unknowingly picked some for the centerpiece of her Thanksgiving table! I'll leave the rest to your imagination). They are deciduous, leaving bare twigs with telltale white berries attached in late fall. It grows anywhere and everywhere—deep shaded woods and open, dry hillsides—it isn't picky. Look for it along low elevation trails in the Trinities like Canyon Creek and the North Fork of the Trinity River. If you find that you have been exposed, wash the affected area with hard soap and COLD water. Hot water only encourages the spread of the oil that causes the itching.

DOG ROSE
Rosa canina
Rose family

THIMBLEBERRY
Rubus parviflorus
Rose family

WESTERN POISON-OAK
Toxicodendron diversilobum
Sumac family

TOYON, CHRISTMAS BERRY, HOLLYWOOD

Heteromeles arbutifolia Alternate leaves - 5 symmetrical petals ovary superior
Rose family

Toyon can grow to 15′ or more with leathery and sharply toothed, oblong leaves. The creamy white flowers bloom in large pyramidal shaped terminal clusters. This is a common shrub of the dry chaparral hillsides throughout this area and brings a bit of visual holiday joy to a drive along any road or highway during late fall and early winter months. I prefer the name Christmas berry to toyon because the bunches of bright red berries always ripen into their prime during the winter season and are used as seasonal decorations. I photographed this bunch along Highway 299 West near Whiskeytown Lake.

CREAMBUSH, ROCK SPIRAEA, OCEANSPRAY, ARROWWOOD

Holodiscus discolor Alternate leaves - 5 symmetrical petals ovary superior
Rose family

Creambush is common throughout the area covered by this book. It's a stiff shrub that grows on steep hillsides, ridgetops, and rocky slopes—often in the crevices of rocky outcrops and cliff faces at mid to high elevations. The beautiful displays of white flowers are what initially attract your eyes. They bloom in thick clusters and have a decidedly fuzzy look because of the numerous and prominent stamens. The deep green leaves are ovate, toothed or lobed at the margins, and serrate with prominent pinnate veins. Native Americans use the straighter stems from this shrub for arrow stock which gives rise to one of the common names - Arrowwood. I photographed these on Mt. Ashland in Southern Oregon and along the trail to Hidden Lake in the Trinity Alps.

OSO BERRY, INDIAN PLUM, INDIAN PEACH

Oemleria cerasiformis Alternate leaves - 5 symmetrical petals ovary superior
Rose family

This loosely branched shrub grows in moist, shaded areas at lower elevations throughout the Klamath Mountains, indeed—throughout much of California. The flowers open in late winter in this area, often in March, and grow in pendulous racemes. Petals are white and emerge from between elongated sepals. Male and female flowers grow on separate plants. The leaves are narrowly oval with the edges slightly rolled under. The fruits, green at first, turn apricot-colored and then a deep blue when ripe and are usually covered with a whitish waxy coating. I photographed this bunch along Clikapudi Trail above Shasta Lake, and the adolescent fruit along the Upper Trinity River. **Note:** The fruit is prized by Native Americans as a food source. They prepare the twigs and bark for teas and for medicinal use.

TOYON, CHRISTMAS BERRY HOLLYWOOD
Heteromeles arbutifolia
Rose family

CREAMBUSH, ROCK SPIRAEA, OCEANSPRAY
Holodiscus discolor
Rose family

OSO BERRY, INDIAN PLUM INDIAN PEACH
Oemleria cerasiformis
Rose family

NINEBARK
Physocarpus capitatus Alternate leaves - 5 symmetrical petals ovary superior
Rose family

In the early 70s, when I first began to record the plants of the Trinity Alps, it was with pencil and paper. Ninebark was the subject of my first drawing. I still have it hanging on my wall and every time I look at it, it brings back pleasant memories. I sat for hours at the small spring about one mile up Long Canyon Trail with my drawing pad and pencils capturing this gorgeous shrub. The rounded clusters of small snow-white flowers bloom at the ends of reddish stems. The palmately lobed leaves resemble maple leaves. The bark is loose and shreddy and Native American lore suggests that if you rub your fingers down the stems 9 pieces of bark will fall off in your hand. Look for it along streams and by springs throughout the area covered by this book. I've found it around Shasta and Trinity Lakes as well as at mid-elevation areas in the surrounding mountains. I photographed these along a stream below Kangaroo Lake in the Trinity Divide.

PACIFIC PLUM, SIERRA PLUM, KLAMATH PLUM
Prunus subcordata Alternate leaves - 5 symmetrical petals ovary superior
Rose family

This small tree can reach heights of 20′ but more often it forms thickets of shrubby bushes 8′ to 10′ tall. It is a lower elevation tree often found along streams and on shaded hillsides below 5,000′ in this area. The branches are rough, crooked, and occasionally spiny. Leaves are ovate to almost round and up to 2″ long. Snow white flowers bloom in a loose cluster with oval petals and prominent stamens. The crab apple-sized reddish fruits mature in late summer but good luck finding any, as they are much favored by wildlife. I photographed these near Shasta Lake but look for them throughout the foothill country of the Trinities especially the Trinity River.

CASCADE MOUNTAIN ASH, MOUNTAIN ASH
Sorbus scopulina Alternate leaves - 5 symmetrical petals ovary superior
Rose family

Hikers are more likely to remember this sprawling shrub by the deep gold of its foliage and brilliant oranges and reds of its fruit in the fall of the year. It decorates entire mountainsides with color in places like Little Bear Lake and is one of the many things that make autumn in the high country so memorable. Individual shrubs can be as tall as 12′ - sometimes more. Masses of small white flowers bloom in flat-topped clusters beginning in late May, giving way to bunches of bright orange to deep red, berry-like pomes usually from late August into November. Leaves are oval, opposite on the stems, with toothed margins and pronounced veins. I photographed the flowers at Eaton Lake in the Russians in mid-September, the fruit at Little Bear Lake in October and the small photograph at Alpine Lake in mid-September–both in the Trinity Alps.

NINEBARK
Physocarpus capitatus
Rose family

PACIFIC PLUM, SIERRA PLUM KLAMATH PLUM
Prunus subcordata
Rose family

CASCADE MOUNTAIN ASH
Sorbus scopulina
Rose family

CHAPARRAL CLEMATIS, PIPESTEM CLEMATIS
Clematis lasiantha
WESTERN VIRGIN'S BOWER, WESTERN WHITE CLEMATIS, CREEK CLEMATIS
Clematis ligusticifolia 4 petals
Buttercup family

Years ago I came across a diary kept by a girl who made trips into Redding with her parents from the small community of Vollmers in the late 1800s to buy supplies for their homestead. In it she often referred to her love of finding what she called "traveler's delight" twining through the brush alongside the road. Depending on the location, what she saw could have been either *Clematis* species–chaparral clematis or western virgin's bower. Chaparral clematis prefers dry, brushy hillsides and full sun while western virgin's bower is found in stream bottoms, or wet areas but also in full sun.

The flowers of both species are a lovely creamy white with prominent yellow stamens and anthers. In late summer as the the flowers mature they are replaced with fuzzy seed heads that are just as interesting and showy as the flowers resembling that prominent head of hair worn by Einstein. Initially the seeds are a golden yellow but later they turn an almost translucent white.

I've photographed the flowers and seedheads of *C. lasiantha* in several locations including along Deadwood Road just outside Lewiston and in the hills above Shasta Lake in May and June. I've photographed the flowers and seedheads of *C. ligusticifolia*at Lowden Ranch just above the Trinity River and above Slattery Pond near Junction City in early May. I photographed the seedheads at Lowden Ranch in late October.

WILD CUCUMBER, COAST MAN-ROOT
Marah oregana 5 petals joined into a bell
Gourd family

This is one BAD cucumber–not cool at all, except to look at. The five-petaled snow-white female flowers form a bell or cup with the petals folding away from the center. A mass of deeply inset lemon-yellow stamens occupy the center of the flower. Male flowers are small but numerous, blooming in bunches all along the length of this climbing vine which can reach lengths of 20′ or more. The large leaves are deeply lobed and, in this area fairly smooth in appearance. On the coast the leaves are a deeper green, thicker and more wrinkled looking. Large round to somewhat oblong glossy fruits begin to appear by late spring. They are light green with deep green stripes running the length and are, to varying degrees covered with prickly appendages. Fruit in this area tends to be smaller than the coastal siblings and not near as prickly. By late summer cucumbers hang all shiny and inviting enticing the unwary to take a big bite. Not a good idea, unless you really don't care about your taste buds. The fruit is extremely bitter! It is common throughout this region from the coastline to the foothills at lower elevations.

CHAPARRAL CLEMATIS
PIPESTEM CLEMATIS
Clematis lasiantha
Buttercup family

WESTERN VIRGIN'S BOWER
CREEK CLEMATIS
Clematis ligusticifolia
Buttercup family

WILD CUCUMBER
Marah oregana
Gourd family

CHAPARRAL FALSE BINDWEED, MODOC MORNING GLORY, WESTERN MORNING GLORY
Calystegia occidentalis ssp. *occidentalis* — 5 symmetrical petals forming a tube
Morning glory family

This large flowers of this morning glory form an almost perfect hexagon. The petals are white and joined at the edges. The centers are a light green with light green streaks that extend outward to the petal tips. The stamens are the same light green with snow white anthers. The leaves are shaped like arrowheads and are deep green and smooth. I shot this photograph along the Dog Creek Road east of Clear Creek Campground in mid-May. I've also photographed them along the Waters Gulch Trail above Shasta Lake in early spring.

SIERRA FALSE BINDWEED, SIERRA MORNING GLORY, WOOLLY MORNING GLORY
Calystegia malacophylla — 5 symmetrical petals forming a tube
Morning glory family

The whole aspect of this plant is "stiff" to the touch. The light olive-green leaves, are more-or-less arrow shaped, small, and covered with a dense mat of stiff hairs. The stems, likewise, are stiff and so covered with those same stiff hairs that they appear almost white. Their growth habit is not at all luxuriant like other morning glories. It is much sparser in appearance and hugs the ground in a sprawling habit preferring open, sunny and rocky locations. The flowers are typical of other morning glorys–about 2" across but sometimes a bit larger. I photographed this one along the trail to Marshy Lakes from Tangle Blue but it is quite common throughout the area and especially along Highway 3 between the Bowerman Ridge intersection and Trinity Center, growing on the cutbanks.

CHAPARRAL FALSE BINDWEED
Calystegia occidentalis
ssp. *occidentalis*
Morning glory family

SIERRA FALSE BINDWEED
Calystegia malacophylla
Morning glory family

Black's Basin in the Trinity Alps: This beautiful but isolated little basin is rarely visited by backpackers who favor a lakeside destination. But wildflower enthusiasts would find it absolutely enchanting. From June through October the surrounding meadows are alive with flowers. Through the years I have photographed over 30 species here. If you don't like crowds this is your kind of destination. The view here is towards Caribou Mountain.

THE DODDERS: The holoparasitic dodders illustrated in this book can all be recognized by the masses of thin orange, rubbery, vine-like "stalks" that spread over and twine tightly about the host plant which can be anything that produces chlorophyll. Tiny appendages called haustoria emerge from these stalks and probe into the cambium of the host plant which allows them to gather moisture and nutrients that sustain their life cycle. Being parasitic, they are incapable of producing these things on their own. Leaves which, in chlorophyll producing plants, convert sunlight into chlorophyll, are reduced to tiny scales on dodders and serve little function. All three of these dodders produce masses of flowers the shapes of which are key or diagnostic to identification. As the flowers mature, they produce seeds that drop to the ground just like non-parasitic wildflowers. When the seeds sprout, they have about five days in which the twining stems can reach a host plant and attach themselves before they die from lack of nutrients.

CALIFORNIA DODDER, CHAPARRAL DODDER
Cuscuta californica
Morning glory family

5 symmetrical petals forming a tube

The flowers of California dodder bloom in huge conglomerate masses, are an almost translucent snow white (about pea sized), and form a stubby bell shape. The petals flare away from the center with sharply pointed tips that stand sharply upright. The five stamens are prominent, reaching beyond the bell and tipped with orange anthers. I actually photographed this in my son's yard in West Redding in early July. Other locations include near Lamoine on the east side of the Trinity Divide. The plants I photographed were using pumpkins and sunflowers as host plants–I am assuming that is not the case near Lamoine.

CANYON DODDER
Cuscuta subinclusa
Morning glory family

5 symmetrical petals forming a tube

The tiny flowers of this species are crowded along the stems and form white tube-shaped elongated bells with sharply pointed petal tips that flare away from the center at almost 90°. The stamens are gathered together at the constricted neck of the bell and topped with elongated orange anthers. I shot these photographs along Trinity Mountain Road on the Trinity Divide in late July. There were several colonies within a 200-yard radius using ceanothus and black oak as host plants.

JEPSON'S DODDER
Cuscuta jepsonii
Morning glory family

5 symmetrical petals forming a tube

My co-author Julie Kierstead and Len Lindstrand (botanist with Sierra Pacific Industries) found this dodder while observing vegetative fire recovery along portions of the Trinity Divide burned in the 2018 Hirz, Carr, and Delta Fires. It is considered rare. The tiny pea-sized flowers bloom in large masses, are rounded globose with petals that open near the tips, and stand nearly erect. They do not flare away as in the other two species illustrated here. They are white to greenish white with orange stamens found down inside the petals–not prominent as with the others. I photographed these in late July on the road to the Slate Mountain Lookout site where they were growing over mahala mat using it as a host plant.

CALIFORNIA DODDER
Cuscuta californica
Morning glory family

CANYON DODDER
Cuscuta subinclusa
Morning glory family

JEPSON'S DODDER
Cuscuta jepsonii
Morning glory family

CUTLEAF BLACKBERRY
Rubus laciniatus
HIMALAYAN BLACKBERRY
Rubus armeniacus
Rose family

5 symmetrical petals ovary superior

How about some blackberry cobbler or blackberry jam? Maybe a warm piece of blackberry pie with vanilla ice cream? Or picked fresh right off the vine and eaten right out of the picking bucket. Any way you eat them, they are absolutely delicious. Picking blackberries is one of those all-American past-times and in these mountains, there is no shortage of the juicy purple-black fruit to enjoy. They grow thick along most area streams, along the edges of lakes and around seeps and springs—any place where there is water. I can't think of anyone who doesn't know what blackberry vines or fruit look like so that being said I don't think they need much in the way of a description.

These are three of the species found in this area - California (*R. ursinus*) which is native and the non-native cutleaf (*R. laciniatus*) and Himalayan (*R. armeniacus*). The fruit of all three is identical but there are differences in the leaf structure, the stems, and in the flowers.

California and Himalayan blackberries are relatively easy to tell apart. The leaves of the Himalayan species are heart shaped, with more finely serrated edges than those of *R. ursinus*, which has more grooves. The five petals of the Himalayan blackberry are generally fuller and wider than the California blackberry, and the thorns are more abundant on the non-native Himalayan blackberry. Flowers of *R. ursinus* start out pink but generally fade to white with a light pink tinge while those of *R. armeniacus* tend to be somewhat pink. The more oval shaped leaves of *R. ursinus* and *R. armeniacus* are similar but cutleaf (*R. laciniatus*) comes by its name honestly because the leaves are deeply dissected, sharply pointed and somewhat folded at the major mid-ribs.

WILD RASPBERRY, BLACKCAP RASPBERRY, WHITEBARK RASPBERRY
Rubus leucodermis
Rose family

5 symmetrical petals - ovary superior

Here is a treat for any woodland forager. The vines grow in profusion from low- to mid-elevations throughout the foothills and mountains of this area. They are especially abundant in disturbed areas alongside roads and trails. If you can beat hungry animals to the fruit you are in for a tastebud delight. No matter where I am and despite the urgency of the situation (short of a medical emergency), I am always sidetracked by the sight of ripe raspberries when I'm hiking and will spend as much time as it takes to get my fill! The spiny vines grow in sprawling masses along the ground and, in May and June (sometimes into July) are covered with snow-white five-petaled flowers typical of the rose family. Berries form soon afterwards and by the time late summer rolls around the tasty fruit is ready for harvest. They get their name "blackcap" raspberry because of their dark color and the fact that when picked they slip off like little beany caps. I photographed the flowers along Trinity Mountain Road near VanNess Gulch on the Trinity Divide in June and the fruit along the trail above Enni Camp on the way to Papoose Lake in the Trinity Alps in late August.

HIMALAYAN BLACKBERRY
Rubus ameniacus
Rose family

CUTLEAF BLACKBERRY
Rubus laciniatus
Rose family

WESTERN RASPBERRY, WHITEBARK RASPBERRY, WHITE STEMMED RASPBERRY
Rubus leucodermis
Rose family

BUCKEYE, HORSE CHESTNUT
Aesculus californica
Soapberry family

4 petals

If there is any shrub or tree more perfectly suited to California's Mediterranean climate, I don't know what it might be. Buckeye comes on gang busters in the spring with beautiful bright green foliage followed, within a short time, by an awesome display of beautiful flowers. Leaves are large, opposite, and palmately lobed. The flowers are creamy white with prominent stamens topped with bright orange anthers and bloom in tall spikes. By mid-summer as temperatures soar the leaves wither, turn brown, and drop to the ground leaving tangles of bare branches from which hang the developing fruit–large golden-brown nuts. These nuts are much favored by wildlife and shortly after dropping to the ground have been largely cleaned up by foraging deer and squirrels. Native Americans used them as a food staple by grinding them into a flour then leaching them in running water to remove tannic acid. This is a foothill species and is common throughout this area. I photographed these near Shasta lake and Lewiston.

BROWN DOGWOOD
Cornus glabrata
Dogwood family

4 petals

For years this beautiful little dogwood was blooming right under my nose in a place where I've launched my canoe and kayak hundreds of times. As attentive as I am to anything flowering this one always escaped my notice. It is a small, somewhat shrubby tree that spreads in clonal thickets. True to its name, flowering stems are brown to sometimes reddish-brown with leaves that are lanceolate to narrowly oval and paler on the underside. Petals are dull white but I find them anything but dull –in fact, they are quite striking against the deep green of the leaves with four prominent copper colored radiating stamens. The ovary is likewise copper colored. I had been searching for this dogwood but it was James Lee, working for the Trinity River Restoration Project, who "discovered" this population and turned me on to it. I took the photographs in early June along the shoreline of Lewiston lake.

PACIFIC DOGWOOD
Cornus nuttallii
Dogwood family

Many petals

If there was ever a reason to take an old-fashioned Sunday drive through these mountains this beautiful tree would be it. From April until early June (and sometimes beyond) the huge "flowers" of this dogwood grace mountain slopes, stream bottoms and otherwise shaded areas en masse, making for an absolutely delightful display. The trees can grow to 30 or 40' in this area, and form an understory of sorts in areas of mixed conifers and hardwoods. Those gorgeous white "petals" are not petals at all but the showy bracts that surround a central cluster of small creamy white flowers. The flowers give way to clusters of bright red, fleshy "berries" in the fall and the leaves turn deep red, almost scarlet, putting on another beautiful display. I took these photographs along Rush Creek in early May. It isn't a frequent occurrence, but I have, in the past, been surprised to find beautiful blooms even in September–especially along the east facing slope of the Russians.

BUCKEYE, HORSE CHESTNUT
Aesculus californica
Soapberry family

BROWN DOGWOOD
Cornus glabrata
Dogwood family

PACIFIC DOGWOOD
Cornus nutallii
Dogwood family

RED OSIER DOGWOOD, CREEK DOGWOOD
Cornus sericea
MINER'S DOGWOOD, BLACKFRUIT DOGWOOD
Cornus sessilis 4 petals
Dogwood family

Both of these dogwoods have their feet in water sometimes forming dense stands next to streams, springs, and the shorelines of lakes. They can appear as shrubs or small trees, multi-stemmed, and growing to near 12' in height. Both are common at mid-elevations throughout this area. The leaves of both species are opposite, with the curved veins characteristic of all dogwoods. Flowers occur in loose, mounded clusters at the ends of the stems. The large white petaloid bracts of their larger cousin Pacific dogwood are absent from these species. To the amateur, they appear similar in habit as well as habitat and, except for the fruit and twigs, are hard to tell apart. The fruits of *sericea* are small and pale greenish-white and its twigs are red (top left inset). The fruit of *sessilis* are large, fleshy, and blue-black (top right inset), and its twigs are brownish or gray. The photographs of *sericea* were taken along the shoreline of West Boulder Lake in the Trinity Alps in mid-July while the inset of the *sessilis* berries was taken along Slate Creek near the Kinney Camp Trailhead in the Trinity Alps in October.

MADRONE
Arbutus menziesii *5 petals, joined into a tube*
Heath family

With its beautiful smooth brownish-red bark, deep green leathery leaves, and myriad clusters of creamy-white shaped flowers, one could almost imagine this as some form of giant manzanita. They do belong to the same family but that's pretty much where the likeness ends. Lower down on the trunk the bark becomes gray-brown and deeply furrowed and that beautiful reddish bark higher up splits in the late spring and falls away much the same way that manzanitas do, leaving the new emerald green bark exposed beneath. In the fall the flowers give way to bright red, fleshy "berries." Overall, it does not need a detailed description because it is definitely one-of-a-kind in these mountains and cannot be confused with any other tree. This large, beautiful tree is found all along the Trinity River and, for wilderness backpackers, along the lower reaches of the Canyon Creek Trail. These photographs came from along the Trinity River and Canyon Creek.

RED OSIER DOGWOOD & MINER'S DOGWOOD
Cornus sericea & *Cornus sessilis*
Dogwood family

MADRONE
Arbutus menziesii
Heath family

Yellow fawn lilies in bloom on the slopes above Horse Creek in the Trinity Alps.

Sulfur-flowered buckwheat - *Eriogonum umbellatum* - in bloom along the Deacon Lee Trail in the Russian Wilderness.

ARROW-LEAF BUCKWHEAT, COMPOSITE BUCKWHEAT, HEARTLEAF BUCKWHEAT
Eriogonum compositum var. *compositum* 3 or 6 petals
Buckwheat family

This is the yellow flowered version of the white arrow-leaf buckwheat shown on page 24. The two are identical in all other ways. It's one of the largest buckwheats in these mountains and easy to identify mainly because of its size–up to 2′ tall in some places. The stems are stout and the basal leaves an elongated heart shape, dark green above, and white and wooly beneath. The tiny, predominantly white flowers bloom in compact heads that sit atop several elongated branches which in turn arise from the same spot at the top of the main stem. This flower arrangement is referred to as an umbel. It likes open rocky areas and meadows. I photographed this one in late June along the Gazelle to Callahan Road where it was blooming on roadside cutbanks and other exposed areas.

CONGDON'S BUCKWHEAT
Eriogonum congdonii 3 or 6 petals
Buckwheat family

Congdon's buckwheat is one of the easiest buckwheats to identify. Unlike many other species in the genus, the branches of the flower clusters (inflorescences) are clearly visible above the small leaf-like bracts. The leaves are small, greenish-wooly above, densely white-wooly below, and curl under along the edges–making them appear narrow. The plants can be up to 12" tall, including the flowers. This species is endemic to serpentine and is found in the Trinity Alps and along the Trinity Divide. It is one of the most common buckwheats in the Mt. Eddy to Scott Mtn. area. I photographed these above Kangaroo Lake in Mid-July.

JAYNE'S CANYON BUCKWHEAT
Eriogonum diclinum 3 or 6 petals
Buckwheat family

This small buckwheat blooms at higher elevations throughout the central Klamath Mountains. This species seems to shun serpentine; it is absent from the Trinity Ultramafic Sheet (Eddys, Scott Mtn., Red Trinities). The plants I've photographed have rarely been over 5″ tall, including the blooms. The compact bunches of basal leaves are mostly egg shaped, light bluish-green, and covered with whitish hairs making them appear soft. Like some of the other buckwheats the flowers of this one are tinged with red later in the summer season. I photographed these above the outlet of Sapphire Lake in late August.

ARROW-LEAF BUCKWHEAT
Eriogonum compositum
var. *compositum*
Buckwheat family

CONGDON'S BUCKWHEAT
Eriogonum congdonii
Buckwheat family

JAYNE'S CANYON BUCKWHEAT
Eriogonum diclinum
Buckwheat family

LOBB'S BUCKWHEAT
Eriogonum lobbii
Buckwheat family

3 or 6 petals

This is a fairly easy buckwheat to identify. The rounded heads of lemon yellow flowers emerge from beneath a basal rosette of leaves and lie flat against the ground. The leaves are broadly oval, fuzzy, with white veins. It grows commonly in granitics, often in the cracks and fissures of rock outcrops. In late summer, in open exposed areas, the leaves and flowers turn a deep and vibrant red. This is especially true on the upper, exposed reaches of the Suicide Trail in the Trinity Alps (see inset).I photographed this one in late August above Upper Albert Lake in the Russians but have also photographed it above Lower Caribou Lake in the Trinity Alps.

NAKED-STEM BUCKWHEAT

3 or 6 petals

Eriogonum nudum
Buckwheat family

Here's a distinctive buckwheat you don't even have to get out of your car to see. It is common alongside roads, highways, and trails everywhere. The plants are tall (up to 2′ or more), multi-branched, and gangly in appearance. The buttery-yellow flowers bloom atop the naked stems in little round clusters–one cluster per stem with many stems per plant. This pretty buckwheat prefers dry, open, and exposed sites. Stems turn red in the fall. I shot these photographs above Trinity Lake near Buckeye Ridge in June.

SISKIYOU BUCKWHEAT
Eriogonum siskiyouense
Buckwheat family

3 or 6 petals

This is a small buckwheat that likes serpentine soils. The lemon yellow flowers bloom in small, tight clusters at the tops of strongly erect stems which arise from dense masses of tiny, almost succulent looking leaves. There is a distinctive single whorl of 3 leaf-like bracts about 2/3 of the way up the stems–visible in the large photograph. Late in the season the flowers take on a rusty appearance but the leaves are the real attraction when they turn a vibrant and deep reddish orange. I photographed these along the Pacific Crest Trail above Kangaroo Lake in mid-July.

LOBB'S BUCKWHEAT
Eriogonum lobbii
Buckwheat family

NAKED STEM BUCKWHEAT
Eriogonum nudum
Buckwheat family

SISKIYOU BUCKWHEAT
Eriogonum siskiyouense
Buckwheat family

SULPHUR-FLOWERED BUCKWHEAT

Eriogonum umbellatum — 3 or 6 petals

Buckwheat family

This is probably the showiest and most recognizable buckwheat anywhere in the high country. It is common and widespread at mid to high elevations throughout these mountains and favors open, dry, exposed sites. The thin leaves are small, basal, and dark green on the top side, light gray-green underneath. It blooms in large colonies with so many flowers that they can color entire hillsides and rocky outcrops as shown in the large photograph. Early in the blooming season the flowers are a distinct lemon yellow but as the seasons progress they begin to turn a reddish-orange and this color persists into the early winter months. I shot the large photograph along the old Bear Creek Trail below Monument Peak in early August. Another gorgeous display can be seen along the Deacon Lee Trail in the Russians. **Note:** There are currently 25 subspecies of this highly variable species known from within California. Only *Eriogonum umbellatum* var. *humistratum* (shown above) is listed as a separate entry here because it is a near endemic to the book's area.

MT. EDDY BUCKWHEAT, SCOTT MOUNTAIN SULPHUR FLOWER

Eriogonum umbellatum var. *humistratum* — 3 or 6 petals

Buckwheat family

The bright yellow flowering heads of this buckwheat are similar to those of the more common sulphur-flowered buckwheats. They bloom in thick, rounded heads atop slender stalks emanating from the basal leaves. The leaves are a distinguishing factor–being fleshy and covered with whitish hairs. Mt. Eddy buckwheat is found on serpentine soils at higher elevations from ranging from Mt. Eddy, north to the Scott Mountains then west into the Trinity Alps, Russians, and Marble Mountains. I photographed these on Scott Mountain in mid-July.

SULPHUR-FLOWERED BUCKWHEAT
Eriogonum umbellatum
Buckwheat family

MT. EDDY BUCKWHEAT
Eriogonum umbellatum var. *humistratum*
Buckwheat family

DWARF OREGON-GRAPE
Berberis nervosa 3 or 6 petals
Barberry family

This is a pretty wildflower found throughout the Klamath in shaded sites–usually beneath a canopy of mixed conifer and sometimes hardwoods–at lower to mid-elevations. Its holly-like leaflets are evergreen, seven or more in number, stiff, and spiny. Although the leaves can change color, as illustrated, they are persistent through winter. It's a low growing plant, usually no more than 18" in height, with woody stems and, in the spring, groups of yellow to yellow-green flowers followed by bunches of dull blue fruit usually coated with a waxy whitish bloom (see inset). There are three species of Oregon-grape found in these mountains but this one is the one you are most likely to encounter alongside trails in the wilderness. The berries in the small photograph actually belong to tall oregon grape (*Mahonia aquifolium*) but look identical to those of *M. nervosa*. I photographed this bunch along Big French Creek in early September where the two can be found in fairly close proximity to one another. I photographed the flowers along the China Springs Trail to Grizzly Lake in August.

YELLOW FAWN LILY, GLACIER LILY
Erythronium grandiflorum subsp. *grandiflorum* 3 or 6 petals
Lily family

This little fawn lily emerges early in the blooming season, often at the edges of melting snow banks and other slightly wet sites in fairly open areas at middle to high elevations. Typically, the bright lemon yellow flowers nod downward from tall 6" to 8" slender stalks. Petals are generally recurved upward exposing prominent whitish stamens, anthers, and stigmas. At higher elevations you will often find this same lily blooming upright on the ends of almost non-existent stalks, nestled tightly into the plain green basal leaves and with only slightly reflexed petals as illustrated in the inset photograph. The large photograph was taken above Horse Creek Meadows and the inset on the ridge between Foster and Lion Lakes, both in the Trinity Alps.

YELLOW BELLS, GOLD BELL, YELLOW FRITILLARY
Fritillaria pudica 3 or 6 petals
Lily family

Through the years I've spent quite a lot of time exploring Scott Mountain and I think I've come to know the area pretty well–yet I come across something new and unexpected with each visit. This charming little fritillary was one of those unexpected discoveries and reaffirms the fact that timing is everything. It was blooming in large but scattered populations in close proximity to melting snow banks. Small, solitary buttery-yellow, bell shaped flowers nod at the ends of stout 4" to 6" stalks. The leaves are basal and tend to be soft grayish-green which blends in with the surrounding exposed serpentine rocks.

DWARF OREGON-GRAPE
Berberis nervosa
Barberry family

YELLOW FAWN LILY
GLACIER LILY
Erythronium grandiflorum subsp. *grandiflorum*
Lily family

YELLOW BELLS
Fritillaria pudica
Lily family

YELLOW-LEAF IRIS, YELLOW-FLOWERED IRIS
Iris chrysophylla
Iris family

3 or 6 petals

This is an iris that, in some instances, could be confused with variations of the long-tubed iris. because both have the same long corolla tube and variable color combinations. The color of this particular iris tends to yellow or a creamy yellowish-white with a somewhat pale bluish tinge. The ones I've photographed along the Stoney Ridge Trail appear in two variations: the more common one having beautiful, soft yellow petals with purple veining and a deeper yellow stripe running the full length of each petal. The other version has mostly creamy white petals with prominent deep purple veining and a yellow stripe running the full length of the petals. To be sure—there are other color variations you might encounter but, generally speaking, this description should suffice. This iris is common and widespread throughout this area.

BOG ASPHODEL, CALIFORNIA BOG ASPHODEL
Narthecium californicum
Bog asphodel family

3 or 6 petals

This is a fairly inconspicuous wildflower found in large populations in wet, marshy areas. I say inconspicuous because it isn't until you get up-close and personal with it that you can truly appreciate the beauty of each tiny flower. The golden-yellow blooms are concentrated along a raceme that can grow to 16" tall but are often hidden in tall grasses. The petals flare away and hang down exposing pretty little wooly filaments topped with deep orange hook shaped anthers. I photographed this one in a wet meadow along the trail to Stoddard Lake. I've also seen it in the meadow at L Lake in the Trinity Alps and in the meadow at the back of Kangaroo Lake on the Trinity Divide. **Note:** In the fall when the flowers go to seed, the seed-heads are every bit as pretty and unusual as the blooms. They look a lot like the little hats artists feel obligated to place on the heads of cartoon gnomes (see inset).

CREAM CUPS
Platystemon californicus
Poppy family

3 or 6 petals

This striking beauty is an early bloomer from the floor of the Central Valley to the lower elevations in the foothills. It is also highly variable in color—ranging from solid yellow to various combinations of yellow and white or orange and white. The flowers sit solitary atop 2½" to 4" stems and grow in profusion on exposed, sunny flats and hillsides. The diamond shaped petals form a flower that can be up to 1" in diameter. I found the beautiful golden-yellow and white version blooming in large colonies together with baby blue eyes up along Canyon Creek Road in mid-April. The mostly white version I photographed along Clikapudi Trail near Shasta Lake in late March.

YELLOW-LEAF IRIS, YELLOW-FLOWERED IRIS
Iris chrysophylla
Iris family

BOG ASPHODEL, CALIFORNIA BOG ASPHODEL
Narthecium californicum
Bog asphodel family

CREAM CUPS
Platystemon californicus
Poppy family

GOLDEN EYED GRASS, YELLOW EYED GRASS, GOLDEN BLUE EYED GRASS
Sisyrinchium californicum
Iris family
3 or 6 petals

This distinctive member of the iris family is commonly found on serpentinized soils. It prefers open, wet meadows or a proximity to water. The plant is tall—about 24″ or —with tough, fibrous, strongly erect stems which are sharply bent a few inches below the flowers. Stems are topped with a solitary bright yellow six-petaled star shaped flower. The veins in the petals, if they show, are usually brown. Stamens are prominent and topped with deep orange anthers. I found and photographed these in a wet meadow just off of Forest Highway 24 near Gumboot Lake in late June.

YELLOW TRITELEIA, TRINITY MOUNTAINS PRETTYFACE
Triteleia crocea
Brodiaea family
3 or 6 petals

The first thing you notice about this gorgeous triteleia is the color. The deep, buttery yellow jumps out no matter the surroundings. Though it does not tend to bloom in large colonies, even singularly it is a stand-out. Each petal has a mid-vein of greenish-brown running the length. The anthers are snow-white. The plants range in height from 10″ to 18″. The leaves are almost as long as the flowering stalk but begin to wither and turn brown as the flowers mature. The number of flowers in each head ranged from five to more than a dozen. I photographed these on a flat alongside the road to the Poison Canyon Trailhead in the Trinity Alps in late May. **Note:** The blue version of *T. crocea* (also called yellow triteleia) is featured in the blue section of this guide (pg. 334).

PRETTY FACE
Triteleia ixioides subsp. *anilina*
Brodiaea family
3 or 6 petals

Pretty face is common throughout the foothills and is found in habitats ranging from dry and exposed to shaded and damp. It is usually under a canopy of mixed hard and softwoods. The deep yellow, to straw-colored flowers occur in loose clusters at the tops of stems that reach 10″ in height, sometimes taller. Each petal has a dark, reddish-brown stripe down the center almost to the tip. Several bright blue anthers emerge upright from sleeve-like structures in the center. Flower buds are elongated and yellow with deep almost black to reddish-brown stripes. I shot this photograph in Squirrel Gulch at the North end of Trinity Lake in June.

GOLDEN EYED GRASS
YELLOW EYED GRASS
Sisyrinchium californicum
Iris family

YELLOW TRITELEIA
TRINITY MOUNTAINS PRETTY FACE
Triteleia crocea
Brodiaea family

PRETTY FACE
Triteleia ixioides subsp. *anilina*
Brodiaea family

AMERICAN WINTER CRESS
Barbarea orthoceras
Mustard family

4 petals

The tall, gangly habit and deep golden yellow flowers are a good give-away for this member of the mustard family. Leaves are deep green, pinnately lobed, long, and narrow—with the terminal lobe much larger and rounded with wavy edges. Its preferred habitat is wet meadows, streambanks, or any number of places where it is damp. The four petals are slightly reflexed away from the centers. It is common at lower elevations throughout these mountains. I photographed these in late June along the Trinity River below Lewiston but have seen it along Stewart's Fork and the Upper Trinity River. **Note**: The leaves in these photographs are from a thistle and not American winter cress.

CONTORTED SUNCUP, PLAINS EVENING PRIMROSE
Camissonia contorta
Evening primrose family

4 petals

Four petals, slightly notched at the tips, sit on top of the inferior ovary, and flare away from the center with prominent stamens and style. The style is quite a bit longer than the stamens and is topped with a ball-like stigma that droops slightly downward. Leaves are slightly hairy, somewhat linear or narrowly oval, and range in color from bluish-green to deep green. The flowering stems are somewhat reddish-green. It is common throughout the West. I found this little golden-yellow beauty blooming in sandy soil in the middle of a well-used trail above the Trinity River. I had never seen it before and I've walked that trail many times. I took this photograph at the Lowden Ranch near Lewiston in mid-May.

BUSH POPPY
Dendromecon rigida
Poppy family

4 petals

For most of the year this shrub is nondescript, with stiff branches and leathery gray-green leaves. But in April and May it explodes with beautiful, large golden yellow, bowl shaped flowers. It is a low elevation shrub that grows in exposed, dry sites, often alongside roads where the plants put on a magnificent display. Seedlings germinate profusely after wildfire. I photographed these along Tom Green Road above French Gulch on the more southerly end of the Trinity Divide. They are common along Highway 299 west of Redding and around Shasta Lake.

AMERICAN WINTERCRESS
Barbarea orthoceras
Mustard family

CONTORTED SUNCUP
PLAINS EVENING PRIMROSE
Camissonia contorta
Evening primrose family

BUSH POPPY
Dendromecon rigida
Poppy family

YELLOW WILLOWHERB
Epilobium luteum
Evening primrose family

4 petals

While common throughout the Pacific Northwest, this unusual fireweed is listed as rare, threatened, or endangered in California. I had never seen it before but my co-author Julie Kierstead noted a population along Parks Creek on the Trinity Divide—which is where I shot these photographs. I call it unusual because the plants in this population form low growing, rounded mounds with their feet in the water. The pale-yellow flowers age to a delicate rose and are tucked into the mounds which ranged in height from 12" to 18." I photographed these in mid-September in an area where temperatures had dropped below freezing at night and the bloom was almost finished. After a search over a fairly large area I was lucky enough to find six flowers to photograph. Typical of fireweeds, the opposite leaves are narrowly oval, lance shaped, but small and about 2" long. Interestingly, the flowers unfold much like little roses and are quite attractive. The sepals enclosing the flowers are rimmed in red. The style extends well beyond the flower tube ending in four distinct and pronounced stigma lobes.

SIERRA WALLFLOWER
Erysimum perenne
Mustard family

4 petals

In late June, during a major thunderstorm, I walked to the top of a low ridge above East Boulder Lake in the Trinity Alps and found this gorgeous bunch of spreading phlox and Sierra wallflower blooming as though they were meant to be together! In the midst of the storm with light snow falling I set up my camera gear and went to work (we wildflower photographers are nothing, if not fearless, in the face of being fried by lightning!). I've photographed this wildflower in many settings, from Shasta Lake to the top of Mt. Ashland but this image is one of my all time favorites—with those gorgeous deep yellow blooms emerging from the middle of a patch of phlox in a lightning storm. This flower is easily recognized by the golden yellow blooms and its size—up to 24" tall and its intense sweet fragrance. It is common throughout the area covered by this book.

CALIFORNIA POPPY
Eschscholzia californica
Poppy family

4 petals

This, the quintessential California wildflower, is the state wildflower. It is easily recognized by its vibrant gold and orange blooms and it decorates roadsides throughout the state as well as myriad other open and sunny places at lower elevations. You won't find it in the high country but it does prosper along Highway 299 west and along the Trinity River. The flowers spring from a mound of soft, grayish green, deeply dissected leaves. Each flower has four petals which form a broad cup. The sepals are united into a cone shaped hat that comes off as the petals expand. The center of the golden yellow flower is a darker orange. I photographed these on a flat between Rush Creek Road and the Trinity River near Lewiston in July.

YELLOW WILLOWHERB
Epilobium luteum
Evening primrose family

SIERRA WALLFLOWER
Erysimum perenne
Mustard family

CALIFORNIA POPPY
Eschscholzia californica
Poppy family

CASTLE LAKE BEDSTRAW
Galium glabrescens subsp. *glabrescens* 4 petals
Madder family

The flowers of this species are tiny—no more than 1/4" across. The four greenish-yellow petals point straight out and form a cross (similar to the flowers of climbing bedstraw also pictured). The most distinctive feature of this small plant is the hairy fruit, which is silvery in the sunshine. This bedstraw is common from Castle Crags north to the Scott Mountains and west to the Trinity Alps, but it is easily confused (even among professional botanists) with the rare species *Galium serpenticum* subsp. *scotticum*, which inhabits the same area. This subspecies prefers serpentine, although it can be found on other substrates as well. I shot these photographs above East Weaver Lake in early July.

CLIMBING BEDSTRAW
Galium porrigens 4 petals
Madder family

Ever wonder where the name "bedstraw" came from? Pioneers harvested huge bundles of this plant to fill or "tick" their mattresses. My great-grandmother told me stories of stuffing bedstraw into her own mattress as a little girl in the Oklahoma Territory in the late 1800s. This almost vine-like plant grows everywhere in lower elevations, often twining up through small bushes or anything else near at hand. The stems and leaves are sticky to the touch and the tiny four-petaled yellowish-green flowers and white berries inconspicuous. I photographed these one along Rush Creek near Lewiston.

DYER'S WOAD
Isatis tinctoria 5 symmetrical petals forming a tube
Mustard family

This is a common, invasive roadside wildflower that does well in those disturbed areas next to Highway 3 above Coffee Creek. It's not native but naturalized from the Caucuses Mountains of central Asia. Its overall habit is tall with gray-green lance shaped leaves climbing from the base upward along the stalk to the base of the flowering branches. Dozens of bright yellow flowers bloom in multi-branched florets in a flat topped spray, followed by dangling black triangular fruits. This is one wildflower you cannot miss mainly because of its sheer size—2′ to 3′ tall. **Note:** For thousands of years people have extracted indigo dye from the leaves of dyer woad. It, chemically, is the same dye extracted from true indigo, just a bit weaker.

HOOKER'S EVENING-PRIMROSE
Oenothera elata subsp. *hirsutissima* 4 petals
Evening-primrose family

Here's a common and widespread wildflower you don't even have to get out of your car (or boat) to see—especially if you drive 299 between Junction City and the Coast, the road to Kangaroo Lake on the Trinity Divide, or along the banks of the Trinity River. It typically blooms mid to late summer at lower elevations throughout these mountains. The tall stalks (to 5′ or more) are crowned with large, lemon yellow flowers in a loose cluster—each flower perched atop a red elongated ovary.. It is quite showy and really can't be mistaken for anything else in this area. I shot the large photograph along the Trinity River near Lewiston in July and the inset along the road to Kangaroo Lake also in July.

CASTLE LAKE BEDSTRAW
Galium glabrescens subsp. *glabrescens*
Madder family

CLIMBING BEDSTRAW
Galium porrigens
Madder family

DYER'S WOAD
Isatis tinctoria
Mustard family

HOOKER'S EVENING-PRIMROSE
Oenothera elata subsp. *hirsutissima*
Evening-primrose family

WESTERN BLADDERPOD
Physaria occidentalis subsp. *occidentalis* 4 petals
Mustard family

Here's a tiny member of the mustard family that I've found blooming in the serpentines on Scott Mountain. The plants are about 4" high and hug the ground. The tightly clustered gray-green leaves are spoon shaped and fuzzy. The golden-yellow flowers are tube shaped with petals that flare away to expose a deep orange center circled with stubby stamens and anthers. The fruits are inflated globes, hence the name bladderpod. If you are looking for them, be careful where you walk because these small plants are inconspicuous. I photographed these in mid-July.

MARSH YELLOW CRESS
Rorippa palustris 4 petals
Mustard family

Look for fairly large populations of this tiny cress in wet and open mid-elevation meadows shortly after the snow melts. It is a small wildflower, only reaching to about 8" in height. It is strongly upright with lance shaped leaves scattered the length of the stem. The small, buttery-yellow flowers bloom in large bunches along the upper ends of the stems, are tube shaped, and flare widely at the ends. I photographed these in the open meadows below the Parker Divide in the Trinity Alps in late May.

Packer's Peak in the Trinity Alps is perfectly situated for epic views not only of the Trinity Alps but the Russian Wilderness, Salmon River country, and the distant coast ranges. The hike to the top is a steep one on a trail built years ago for horseback access by Forest Service personnel travelling between Coffee Creek and the Salmon River Ranger Stations. It also provided for service to the lookout tower located on the peak. The trail is little used these days and unrecognizeable in some spots over the steep open slopes. I've been to the top many times—sometimes on snowy days and sometimes when it's toasty but, either way, I find it rewarding not only for the views but for the wildflowers along the way,

WESTERN BLADDERPOD
Physaria occidentalis
subsp. *occidentalis*
Mustard family

MARSH YELLOW CRESS
Rorippa palustris
Mustard family

Looking north east to the Russian Wilderness from Packer's Peak in the Trinity Alps

MOUNTAIN JEWELFLOWER, SHIELDPLANT
Streptanthus tortuosus 4 petals
Mustard family

This is a "get-down-on-your-hands-and-knees-and-take-a-good-look" wildflower. It is also one of that can be confusing to amateurs trying to classify by color. They can range through several colors—some are bright yellow, cream, or orange while others are a muted greenish-purple or deep purple with variations in between. The leaves that encircle the flowering stems generally emulate the color of the flowers although they can be muted with green. The flowers are crowded alternately at the tops of the stalks. When they go to seed, dozens of elongated, lance shaped pods hang down which, together with the leaves, gives the plant its name—shield plant. I shot the top left photograph above Doe Lake, the top right photograph in Horse Creek Meadows, the lower left photograph on Scott Mountain, and the leaves with the deep green pods on the slopes above Horseshoe Lake—all in the Trinity Alps.

MOUNTAIN JEWELFLOWER, SHIELDPLANT
Streptanthus tortuosus
Mustard family

Photograph locations clockwise from top left: Doe Lake in August, Horse Creek Meadows in July, Sco Mountain in July, Horseshoe Lake in September

NEVADA BIRD'S FOOT TREFOIL, SIERRA LOTUS

Acmispon nevadensis

5 irregular petals

Pea family

This member of the pea family forms low growing mats in open, dry, and sunny areas along the road to the Canyon Creek Trailhead–the only area in which I've found it. Five to twelve flowers bloom in compact clusters and are a bright buttery yellow. I have found mats of these plants three feet or more in circumference and absolutely covered with blooms making it hard to miss. Stems and leaves are blue or gray-green and covered with silvery hairs which make it soft to the touch. This is a native western species found from the Pacific Northwest into Baja California.

MEADOW BIRD'S FOOT TREFOIL

Hosackia pinnata

5 irregular petals

Pea family

The thick, rounded clusters of contrasting white, deep yellow blossoms and occasional vibrant red sepals make this lotus readily identifiable. The plants are between 5" and 16" tall and are common alongside low elevation trails where there is plenty of moisture. I photographed these along the road to the Stoddard Lake Trailhead in June.

SNUB PEA, SULPHUR PEA

Lathyrus sulphureus

5 irregular petals

Pea family

The flowers of this pea present quite a contrast and another of those species that is impossible to classify by color. Are they orange or white or do the faint yellow markings place them here in the yellow category? Your guess is as good as mine but here they shall reside. Before they have fully opened the backs of the petals are a beautiful rusty orange while the mature, fully opened flowers, are white with brilliant lavender veins on the banner. This is a typical climbing pea and is found at lower elevations throughout the area, generally in the front country of the Trinities. It prefers forested areas with a mix of sun and shade and likes to climb through brush. I shot these near Shasta Lake but they are found all over.

BIRD'S FOOT TREFOIL, BUTTER 'N EGGS

Lotus corniculatus

5 irregular petals

Pea family

You can find this pretty little non-native, weedy immigrant from Eurasia almost anywhere you go–from the coast to the inland foothills. It grows low to the ground in open, sunny areas and is common along the edges of trails and roads. The bright masses of golden-yellow flowers make it easy to spot and identify. I photographed these along the Trinity River in late June.

NEVADA BIRD'S FOOT TREFOIL, SIERRA LOTUS
Acmispon nevadensis
Pea family

MEADOW BIRD'S FOOT TREFOIL
Hosackia pinnata
Pea family

SNUB PEA, SULPHUR PEA
Lathyrus sulphureus
Pea family

BIRD'S FOOT TREFOIL, BUTTER 'N EGGS
Lotus corniculatus
Pea family

MT. EDDY LUPINE
Lupinus croceus
Pea family

5 irregular petals

The compact buttery-yellow flowers on this multi-stemmed lupine make it easy to identify and hard to miss. That being said, people occasionally mistake it for yellow false lupine but Mount Eddy lupine has more than three leaflets per palmately compound leaf while yellow false-lupine has only three (like clovers). This wildflower can grow to 18" and often blooms in large colonies. These photographs were taken along the Stony Ridge Trail and at Stonewall Pass in the Trinity Alps but this species is widespread throughout the mountains and a strong indicator of serpentine soils.. It is common along most roadsides and can add a splash of color to any Sunday drive.

YELLOW FALSE-LUPINE
Thermopsis gracilis
Pea family

5 irregular petals

It's easy to understand why, from a distance, some people confuse this non-lupine with Mt. Eddy lupine. Check it out up close, however, and the differences become apparent. These flowers are a pretty lemon-yellow and loosely clustered along tall sometimes nodding stalks. The leaves have only three leaflets per compound leaf and are somewhat larger than those of Mt. Eddy lupine and almost droopy in habit. They usually bloom in large populations but finding individual plants is not uncommon. They also tend to bloom at lower elevations. I have photographed them above Shasta and Whiskeytown Lakes but I shot these photographs along the lower part of the Boulder Creek Trail in the Trinity Alps in late June. They are common throughout these mountains

STREAM VIOLET, SMOOTH YELLOW VIOLET, PIONEER VIOLET
Viola glabella
Violet family

5 irregular petals

This violet likes shade and water and, true to its name, grows along streams, near springs, and seeps from low to mid-elevations. The heart shaped leaves are slightly cupped, finely toothed at the margins, and deep green. One to three flowering stalks arise from the rootstock bearing one lemon-yellow to buttery-yellow flower in each leaf axil. The lower and two flanking petals are veined with deep purple and the flanking petals each have a slight yellow beard at the opening of the throat. *Viola glabella* can sometimes be confused with *Viola lobata* subsp. *integrifolia* because the leaf structure can appear similar. Look at the back of the flower—if it's yellow it's *V. glabella* (see the small photograph), if it is reddish or purplish it is *V. lobata* subsp. *integrifolia*. I took this photograph along the North Fork of Coffee Creek Trail in the Trinity Alps in early June.

MT. EDDY LUPINE
Lupinus croceus
Pea family

YELLOW FALSE-LUPINE
Thermopsis gracilis
Pea family

STREAM VIOLET
SMOOTH YELLOW VIOLET
Viola glabella
Violet family

DELTA-LEAVED FOREST VIOLET, PINE VIOLET, YELLOW VIOLET
Viola lobata subsp. *integrifolia* 5 irregular petals
Violet family

Two subspecies of *V. Lobata* are illustrated in this book (see below). To distinguish between the two, look closely at the backs of the flowers and the leaves. This one has petal backs that are deep, rusty-red or sometimes purple. Flower color ranges from bright lemon to a soft buttery-yellow. The lower petal is veined with purple in the throat. The two flanking petals have a lemon-yellow beard. The bright green leaves are sharply triangular and mostly entire or shallowly lobed. Look for this one in dry, semi-open conifer or mixed forest stands at low to mid elevations. Both subspecies are common in this area and tend to grow in fairly large colonies, occassionally mixed together. I shot this photograph along the Stewart's Fork Trail in late May.

MOOSEHORN VIOLET, PINE VIOLET
Viola lobata subsp. *lobata* 5 irregular petals
Violet family

Look for this violet in dry, semi-open conifer or mixed forest stands at low to mid elevations. The flowers range from bright lemon to a soft buttery-yellow. The lower petal is veined with purple in the throat. The two flanking petals have a lemon-yellow beard. The backs of the top two petals are yellow. Leaves are gray-green, rounded, and deeply cleft—giving it the common name "moosehorn." The two subspecies are common and tend to grow in fairly large colonies, often mixed together. I shot this photograph near Big Carmen Lake on the Trinity Divide in mid-June.

ASTORIA VIOLET, UPLAND YELLOW VIOLET, CANARY VIOLET
Viola praemorsa subsp. *linguifolia* 5 irregular petals
Violet family

Here is another beautiful little violet that deserves close attention. The flowers look much like other yellow violets with maroon or deep reddish stripes emerging from the throat of the lower three petals and a yellow "beard" at the throat of the two flanking petals. The upper two petals are reddish or brownish-purple on the backs. A single flower blooms at the top of a long, upright stem. The lance or oval shaped leaves are the interesting part—they can be either glabrous (smooth) or covered in dense, silvery hairs. The tips of the leaves are either pointed or rounded. I photographed these near Shasta Lake in April. **Note:** *Viola praemorsa* subsp. *praemorsa* is also found throughout the Klamath Mountains but it is unlikely that the amateur could distinguish between the two.

SHELTON'S VIOLET, CUT-LEAF VIOLET
Viola sheltonii 5 irregular petals
Violet family

Shelton's violet is probably the earliest violet to bloom in these mountains. I have frequently found the plants emerging along lower elevation trails in late February. In most years the flowers appear around Shasta Lake in March. In mid-May they begin to make their appearance in mid-elevation forests. With its almost round and deeply divided blue-green leaves it is unmistakable. The flowers are bright yellow, with deep purple veining extending from the throat—especially on the lower petal. A sparse beard grows at the opening to the throat on the two flanking petals. The two top petals are purple on the backs. The ovary, deep inside the throat, is orange. I photographed this one along the North Fork of Coffee Creek Trail in the Trinity Alps in mid-May.

DELTA-LEAVED FOREST VIOLET, PINE VIOLET
Viola lobata subsp. *integrifolia*
Violet family

MOOSEHORN VIOLET, PINE VIOLET
Viola lobata subsp. *lobata*
Violet family

ASTORIA VIOLET, UPLAND YELLOW VIOLET
Viola praemorsa subsp. *linguifolia*
Violet family

SHELTON'S VIOLET, CUT-LEAF VIOLET
Viola sheltonii
Violet family

MOUNTAIN VIOLET, GOOSEFOOT VIOLET

Viola purpurea 5 irregular petals
Violet family

Of all the violets family found in this area, none are more widely distributed, or as variable in appearance, as the mountain or goosefoot violet. For that reason, I've included photographs and descriptions of the four variations I've found to be most common.

Top left: I photographed this mountain violet in Whiskeytown National Recreation Area in early May. As you will note, the deep green leaves are more or less heart shaped, deeply veined, with toothed margins. Several flowering stems can arise from a single rootstock. The flowers are a beautiful golden-yellow with dark purple veining extending from the throat on the lower petal. The two flanking petals have a sparse yellow beard just inside the throat. All of the petals are purple on the back. Look for this one on open, rocky flats in full sun, or partial shade.

Top right: I photographed this one on the Stoney Ridge Trail in the Trinity Alps in late May. There were large colonies blooming at about 5000′ in shaded sites all along the trail. Typically, the leaves are olive-green, lance shaped, deeply veined, and lightly scalloped at the margins. The lemon-yellow flowers had a rounded look, with shortened petals veined with deep reddish-purple at the throat, and a sparse yellow beard just inside the throat of the two flanking petals. The petals are purple on the backs.

Bottom left: I found this tiny mountain violet along the upper reaches of the Lady Gulch Trail near the divide between Coffee Creek and the Salmon River. The dark green leaves are about 1″ - 1¼″ in diameter, shallowly scalloped, deeply veined, and shaped like the webbed foot of a goose (giving rise to its other common name). The flowers are a deep, buttery-yellow, with reddish-purple veins extending from the throat of the lower petal, and a yellow beard just inside the throat of the two flanking petals. The plants, including the flowering stalk were typically less than 2″ tall and growing in a large colony. I've also found them along the trail to Parker Divide in the Trinity Alps.

Bottom right: This photograph, taken along the Stewart's Fork Trail in the Trinity Alps, represents what one will more typically find at low- to mid-elevations in these mountains. The plants range in height from 4″ to 7,″ with bright green, heart shaped leaves that are broadly scalloped at the margins. The flowers are lemon-yellow with purple veining extending from the throat—more pronounced on the lower petal. Each flanking petal sports a lemon-yellow beard at the opening of the throat. This one likes damp, shaded sites and is commonly found alongside low- to mid-elevation trails in the backcountry.

Whiskeytown National Recreation Area

Stoney Ridge Trail in the Trinity Alps

Whorehouse Gulch (Lady Gulch) - Trinity Alps

Stewart's Fork Trail in the Trinity Alps

MOUNTAIN VIOLET, GOOSEFOOT VIOLET
Viola purpurea
Violet family

SHRUBBY CINQUEFOIL

Dasiphora fruticosa 5 symmetrical petals - ovary superior
Rose family

This cinquefoil is fairly common along the east side of the Trinity Alps and northeast along the Trinity Divide. It is a shrubby plant growing to about 2′ in height. Small tufts of deeply divided lance shaped and somewhat fleshy leaves are scattered along the flowering stalks. The brilliant yellow flowers bloom at the ends of branches or in the leaf axils. The paper thin petals are almost round and surround a crown of deep orange stamens and spatulate shaped anthers in the center. I photographed these along the Fen Trail above Kangaroo Lake on the Trinity Divide in mid-July.

STICKY CINQUEFOIL

Drymocallis glandulosa var. *glandulosa* 5 symmetrical petals - ovary superior
Rose family

The buttery-yellow flower petals fade to a light cream at the tips, and the deep orange stamens in the center are what initially attract you to this beautiful little cinquefoil. The large, heart shaped anthers are another interesting feature as is the pale green calyx that shows between the petals. Unlike shrubby cinquefoil,or the common *Potentilla gracilis* described on page 194,this species has pinnately compound leaves. It is a slender, much branched plant about 12″ tall which likes open, meadows and slopes. I photographed these in the meadows below Long Gulch Lake in early August but you can find them throughout the high country at mid-elevations. I have found them still in bloom in late September around Alpine Lake in the Trinity Alps. The white version of this species is on page 62.

TINKER'S PENNY

Hypericum anagalloides 5 symmetrical petals - ovary superior
St. John's Wort family

Tinker's penny is most often found growing in small colonies alongside streams, the shorelines of lakes and in wet meadows, almost always in full sun. Leaves are rounded, opposite and clasp the short stems from top to bottom. Unlike our other local St. John's worts, the stems are prostrate and mat-forming. Numerous deep yellow flowers about ½″ in diameter with prominent stamens open on each stem. I photographed these at the outlet to Sapphire Lake in the Trinity Alps in September where they were hiding down in tall grass. It is a fairly inconspicuous plant that most hikers discover by accident.

KLAMATH WEED, COMMON ST. JOHN'S WORT

Hypericum perforatum 5 symmetrical petals ovary superior
St. John's Wort family
Klamath weed is a noxious, invasive weed found everywhere in this area. It is especially abundant alongside roads and trails and in open, disturbed areas from the coast to mid-elevations in the backcountry. It is poisonous to livestock and the subject of various eradication efforts through the years. It has been somewhat controlled by a small beetle imported years ago for just that purpose. It is a gangly, leggy, loosely branched plant with opposite stems and small, sharply oval leaves. It can range from a few inches to 3′ tall. The yellow flowers have five petals which are sharply oval or lance shaped with numerous black spots along the edges. The stamens are numerous and prominent. It was used many years ago during the Feast of St. John to ward off evil spirits and has also been used as a treatment for depression—something the folks trying to eradicate it might find useful! I photographed this plant along the road to the Stewart's Fork Trailhead in mid-June.

SHRUBBY CINQUEFOIL
Dasiphora fruticosa
Rose family

STICKY CINQUEFOIL
Drymocallis glandulosa
Rose family

TINKER'S PENNY
Hypericum anagalloides
St. John's Wort family

KLAMATH WEED, COMMON ST. JOHN'S WORT
Hypericum perforatum
St. John's Wort family

ALPINE MOUSETAIL, GORDON'S MOUSETAIL
Ivesia gordonii var. *ursinorum* — 5 symmetrical petals - ovary superior
Rose family

I was sitting in camp at Summit Lake one afternoon with my binoculars glassing the slopes above the lake when this unusual splash of yellow caught my eye. I grabbed my camera and tripod and climbed around the outlet of the lake and up the slope to find this beautiful *Ivesia* in full bloom—growing in the jumbled rocks of the scree slope. It was the first time I'd ever seen it, but since then I've found it blooming in crevices of the cliff face below the outlet of Lake Anna as well. It is characterized by heads of deep yellow flowers and thick bunches of fleshy, dark green, deeply cleft leaves. In this area I've only found it growing on open, exposed serpentine sites but it is not fussy about rock type and is the most widespread of the *Ivesias*. It is found growing in high elevation sites on most mountain ranges in the West. I photographed this bunch in a rocky cleft just below Lake Anna in July.

BREWER'S CINQUEFOIL
Potentilla breweri — 5 symmetrical petals - ovary superior
Rose family

This "leggy" cinquefoil has tall flowering stalks that are divided further into a number of loosely arranged stalks at the ends of which the flowers emerge. The leaves are basal, deep green, pinnately divided and covered with tiny white hairs. A few small leaves bracts are found where the different flowering stalks divide from the main stalk. Flowers in the population I found varied slightly in color from buttery yellow to pale yellow with petals arranged so that the calyx shows through. Stamens, as with the other cinquefoils highlighted in this guide, are prominent with broad, flattened yellow anthers. I took these photographs in mid-June on Scott Mountain where they were blooming all through the open meadows.

NORTHWEST CINQUEFOIL, SLENDER CINQUEFOIL
Potentilla gracilis — 5 symmetrical petals - ovary superior
Rose family

Wander around through the meadows near the Scott Mountain Campground and you will encounter a delightful array of bright yellow cinquefoils including this one. Initially it's hard to tell them apart because the flowers are almost identical—but look to the leaves. *Potentilla gracilis* has palmately compound leaves, unlike the pinnately compound leaves of *P. breweri*. The flowers bloom individually on a tall, somewhat spindly stalk, are a striking yellow, and cup-shaped with broad petals that are ruffled at the tips. The numerous and prominent stamens are yellow topped with yellow and brown anthers. I photographed these in mid-June on Scott Mountain.

WESTERN BUTTERCUP
Ranunculus occidentalis — 5 symmetrical petals - ovary superior
Buttercup family

This tall (12"- 20") and loosely branched wildflower is common at lower elevations—seeking out open areas with lots of sun. It sometimes begins its bloom in February in the foothills around the Sacramento Valley and in March in the foothills surrounding the south and east facing slopes of the Trinity Alps. Like its cousin the alpine buttercup, this one is easily identified by its glossy yellow flowers that appear to have been waxed! I photographed these on Baxter Ridge juist off Highway 3 above Trinity Lake in early June.

ALPINE MOUSETAIL, GORDON'S MOUSETAIL
Ivesia gordonii var. *ursinorum*
Rose family

BREWER'S CINQUEFOIL
Potentilla breweri
Rose family

NORTHWEST CINQUEFOIL
Potentilla gracillis
Rose family

WESTERN BUTTERCUP
Ranunculus occidentalis
Buttercup family

THE EDDYS STONECROP
Sedum kiersteadiae 5 symmetrical petals - ovary superior
Stonecrop family

Stonecrops are difficult to identify but this one, along with the others chosen for this guide, should make the task easier. The Eddys stonecrop, has numerous flowers blooming in a loose bunch on an upright stalk They are bright yellow, with sharply pointed petals streaked with red down the center. When fully opened, the petals flare away at right angles from a reddish-orange ovary and stamens. The backsides of the petals are also reddish-orange becoming more yellow towards the tips and edges. The basal rosette of leaves as well as the stems are succulent looking, fleshy, and a deep green to maroon. Smaller leaves scattered along the flowering stem appear grayish-purple. This photograph was taken above Caribou Lake in the Trinity Alps in mid-August. **Note:** This pretty sedum was named after Julie Kierstead–an expert in the field of stonecrops and one of the editors of this guide.

LANCE-LEAF STONECROP
Sedum lanceolatum 5 symmetrical petals - ovary superior
Stonecrop family

Of all the stonecrops this just might be the easiest to identify. It is quite striking and blooms in abundance on open serpentine slopes. It is is characterized by the reddish, succulent looking leaves and deep yellow flowers tinged with red. Most of the flowering stalks are short with the flowers nestled in a compact bunch just above the leaves. I photographed these in early July on the slopes of Scott Mountain. On a recent pack trip into East Boulder Lake I discovered a huge population of them emerging along the north side of the lake and along the trail to Mill Creek Lake.

BROADLEAF STONECROP, PACIFIC SEDUM
Sedum spathulifolium 5 symmetrical petals - ovary superior
Stonecrop family

Look for this member of the stonecrop family at lower elevations throughout the region–alongside back roads on shaded, damp, steep, rocky hillsides. The perfect little flattened rosettes of succulent leaves colonize large areas when conditions are right. The beautiful little deep yellow flowers bloom in widely branched bunches on stalks that sometimes reach 8" to 10' in height. I photographed these along the road to the Canyon Creek Trailhead in late May.

THE EDDYS STONECROP
Sedum kiersteadiae
Stonecrop family

LANCE-LEAF STONECROP
Sedum lanceolatum
Stonecrop family

BROADLEAF STONECROP
Sedum spathulifolium
Stonecrop family

CREEPING SIBBALDIA, SIBBALDIA
Sibbaldia procumbens 5 symmetrical petals - ovary superior
Rose family

This little wildflower is a world traveller, found throughout the northern latitudes on tundra and in alpine and subalpine regions. I have photographed this species in the Bernese Alps in Switzerland. It forms multi-stemmed, low growing mats in open, gravelly areas. The grayish-green leaves are divided into three wedge shaped leaflets which have three teeth at the tips. Several flowering stalks arise from each plant and each hosts bunches of bright yellow flowers surrounded by hairy green bracts. I found and photographed this bunch in an open, exposed area just above Upper Caribou Lake in late August. It was common throughout the area.

Upper Caribou Lake in the Trinity Alps offers up some magnificent scenery to those willing to make the trek and LOTS of people do. It can be crowded, especially over the major holiday weekends with backpackers and horsepacking groups alike. There are four lakes in this basin beginning with Snowslide, Lower Caribou, Little Caribou, and Big or Upper Caribou–each one is a prize worth seeing at least once. Upper Caribou is the largest lake in the Trinity Alps. This basin is home to myriad wildflowers which is a major attraction for me and my camera. My first visit here was in 1957 on a dayhike over the old Caribou Trail with my flyfishing parents who heard tales of the monster trout found in the upper lake. They were not disappointed but it made for a long day–especially when you consider that in those days all of the roads between Lewiston and Big Flat were dirt and narrow. We left home at 3:00 in the morning and got home at 3:00 in the morning!

CREEPING SIBBALDIA
SIBBALDIA
Sibbaldia procumbens
Rose family

Upper Caribou Lake

TOWERING LOUSEWORT
Pedicularis bracteosa var. *flavida* 5 irregular petals forming a tube
Broomrape family

This is the giant amont the local louseworts---often reaching heights of 4′ or more with flowers crowded along the upper fourth of the stalk. This species' variety is endemic to the moist upper montane conifer forests of the Klamath Mountains. My grand-daughter insists the individual flowers look like miniature dolphins and not their namesake lice. I think she is right. Look for them in marshy areas or wet meadows often growing with corn lilies. I photographed these along the shoreline of Upper Albert Lake in the Russians in late September.

CALIFORNIA STONESEED, CALIFORNIA PUCCOON, CALIFORNIA GROMWELL
Lithospermum californicum 5 symmetrical petals forming a tube
Borage family

This is a common native plant that grows in profusion along the road to the Canyon Creek Trailhead from late May well into June. The plants are fairly tall, reaching 12″ or so, and the leaves and stems are covered with soft hairs. The deep buttery-yellow flowers are tube shaped and about 3/8″ in diameter. Up close I find the plant very attractive and I like the name puccoon (say that 50 times very fast!). Always curious about the origins of the common names, I checked out "puccoon" and found that it was derived from the indigenous Powhatan word "poughkone" in reference to this group of plants, especially the roots, which have been used for centuries to produce dyes. I photographed these along the road to the Canyon Creek trailhead in May

AROMATIC SPRING-PARSLEY, NORTHERN INDIAN PARSNIP, TURPENTINE CYMOPTERUS
Cymopterus terebinthinus 5 symmetrical petals ovary inferior
Carrot family

This pretty and gregarious plant is common at lower elevations in the area covered by this guide. It is often found covering large sections of road-side slopes with its beautiful deep green fern-like foliage and large umbels, or umbrella-shaped clusters of bright yellow flowers. The flowers bloom at the tops of thick stalks that often reach 2′ in height or more. The plant has a decidedly strong anise smell that I find quite pleasant. I photographed these beautiful plants in full bloom above the McCloud River along Gilman Road in early April but look for them everywhere.This looks a lot like a lomatium; the difference is that lomatium fruits are winged on two sides, while cymopterus fruits are winged all around. Fruits are so important to Apiaceae ID!

WOOLY FRUITED LOMATIUM, WOOLYFRUIT DESERT PARSLEY
Lomatium dasycarpum var. *dasycarpum* 5 symmetrical petals ovary inferior
Carrot family

This lomatium favors serpentine soils and can be found throughout areas where this substrate is common. The gray-green branches spread outward from a deep taproot, form a dense ground cover, and are slightly hairy or tomentose. Leaves are likewise gray-green, deeply divided, and cupped. Yellow flowers bloom in dense heads. Stamens are prominent. The foliage, when rubbed, has a distinctive carrot smell that I find quite pleasant. I photographed this one at higher elevations on the Stony Ridge Trail in early June.

TOWERING LOUSEWORT
Pedicularis bracteosa var flavida
Broomrape family

CALIFORNIA STONESEED, CALIFORNIA PUCCOON
Lithospermum californicum
Borage family

AROMATIC SPRING-PARSLEY
Cymopterus terebinthinus
Carrot family

WOOLY FRUIT DESERT PARSLEY
Lomatium dasycarpum var. *dasycarpum*
Carrot family

FOOTHILL LOMATIUM, SPRING GOLD
Lomatium utriculatum 5 symmetrical petals - ovary inferior
Carrot family

Walk any open, sunny flat, or hillside in the foothills in early spring and you will find spring gold in full bloom. It offers up splashes of bright color in areas still recovering from the effects of winter. It is common and unmistakable. Being a member of the carrot family, it has leaves that are finely dissected and do smell a bit like carrot leaves when crushed. The small bright yellow flowers bloom en masse in flat-topped clusters on top of stout stems. The entire plant rarely exceeds 6″ or 7″ in height. I photographed this one blooming on Bolt's Hill above Lewiston Lake in late May.

BLAZING STAR
Mentzelia laevicaulis var. *acuminata* 5 symmetrical petals ovary inferior
Blazing star or Loasa family

This 3′ to 4′ tall, gangly plant decorates roadsides and disturbed areas at lower elevations throughout the area. Considering its large size, it remains relatively unassuming until it blooms—at which time it is a show stopper with dozens of large, sunshine-yellow flowers on each plant. The slender, tapered petals flare away from a host of tall stamens when fully opened. The stems are a light gray and the heavily toothed leaves are a dull olive-green. Look for it in mid-summer along Highway 299 between Weaverville and Big Bar or the Gazelle-Callahan Road as it descends into the Scott Valley.

NORTHERN SANICLE, SIERRA SANICLE
Sanicula graveolens 5 symmetrical petals - ovary inferior
Carrot family

Photographing wildflowers on a bare ridgetop during a raging thunderstorm is probably not the smartest thing to do, but when the opportunity arises to shoot something new do you pass up the chance and run for cover? For me the answer is no. Fortunately I've never been fried by lightning—close on occassion—but not this day! Returning to camp from a day hike to Lion Lake in the Trinity Alps I was attracted by a distant splash of muted color uphill from the trail and that is why this picture is important to me! I barely had time to make one exposure before the storm broke and I was being battered by hail and gale force winds. Fortunately, the exposure was good and the detail excellent. There were several of these little bunches spread out along the ridge—all about 6″ tall with rounded bunches of pale yellowish-green flowers, each bunch on a separate purple-brown stalk. The deep green leaves were rimmed with purple and deeply cleft.

HOWELL'S TAUSCHIA
Tauschia howellii 5 symmetrical petals - ovary inferior
Carrot family

I took this photograph while working with Julie Knorr documenting rare plants for the Klamath National Forest on the Siskiyou Crest in Siskiyou County. This is the only time I've ever seen it. It has also been found in the Marble Mountains but not, as far as I know, anywhere else. It prefers dry, open sites with gravelly granitic soils at higher elevations. The entire plant is less than 4″ tall, with deeply dissected, spiny leaves that are taller than the flowering stems. The flowers are generally yellow and occur in loose clusters surrounded by leafy bracts. There are two other *Tauschias* that could be mistaken for *T. howellii.* They are *T. glauca* and *T. kelloggii*, but they are typically much larger, being 4″ to 20″ in height with flowers occurring in tighter bundles. If Howell's tauschia is going to be found anywhere outside the Marbles in this area I would suspect that it would be along the high ridges of the northwestern portion of the Trinity Alps or along the spine of the Russians.

FOOTHILL LOMATIUM, SPRING GOLD
Lomatium utriculatum
Carrot family

BLAZING STAR
Mentzelia laevicaulis var. *acuminata*
Blazing star or Loasa family

NORTHERN SANICLE, SIERRA SANICLE
Sanicula graveolens
Carrot family

HOWELL'S TAUSCHIA
Tauschia howellii
Carrot family

COBWEBBY INDIAN PAINTBRUSH
Castilleja arachnoidea
Broomrape family

5 irregular petals forming a tube

This is an interesting paintbrush because colors are somewhat variable in these mountains. They can range from a soft yellow to a distinctive rusty-orange as shown. The stems always look like they are covered in white cobwebs. Overall, everything is hairy–including the petals and leaves. Because the yellow form is muted, it's easy to walk past and not really notice them—especially if you're straining under the weight of a fully loaded pack. Overall they are about 12″ tall, multi-branched, with stiff gray-green leaves. The photograph of the orangish flowers was taken near Tangle Blue Lake and the yellow ones on the left in Doe Flat–both in the Trinity Alps in late August. All forms are common and widespread in the Klamath Mountains.

FIELD OWL'S-CLOVER
Castilleja campestris subsp. *campestris*
Broomrape family

5 irregular petals forming a tube

This tiny owl's-clover rarely gets over 3″ tall and, because of its size, hides nicely in open, dry, vernally damp grassy meadows. The inflated, lemon-yellow flowers (about ¼″ across) form a tube with the three-lobed lower lip hanging down in a fan shape from the red spotted throat. The stems, leaves, and sepals are all fuzzy. I photographed these along the Kanaka Peak Trail in Whiskeytown National Recreation Area but I've also seen them along the lower stretches of the East Boulder Lake trail in the Trinity Alps.

CHICKWEED MONKEYFLOWER
Erythranthe alsinoides
Lopseed family

5 irregular petals forming a tube

Look for crowded colonies of this tiny monkeyflower in rocky clefts near, seeps, and the edges of wet meadows throughout these mountains. They are widespread and common. It makes up for its size with sheer numbers and is one of the small wildflower that you are unlikely to miss. The plants range anywhere from 3″ to 6″ in height with flowers that are a beautiful buttery-yellow and about ¼″ in diameter. Typically, you will find a few tiny red spots and one large red central spot on the lower petal. Occasionally, however, you will find them with no red spots at all or a solitary large red spot. I shot the large photograph along the Canyon Creek Trail above The Sinks in June and the small photograph of the flowers without spots in the limestones around the McCloud arm of Shasta Lake in early May.

GIANT MONKEYFLOWER, COMMON MONKEYFLOWER, SEEP MONKEYFLOWER
Erythranthe guttata
Lopseed family

5 irregular petals forming a tube

Erythranthe guttata is a huge complex but the giant, or common, monkeyflower is the one most people are familiar with. It is common in this area and found from Pacific Ocean beaches to wet areas on the floor of the Central Valley and everywhere in-between. The large, buttery-yellow flowers congregate in large groups at the ends of the flowering stalks and have small red spots on the lower corolla lip. The stems and leaves are hairy and somewhat sticky. They bloom prolifically wherever slow flowing or standing water is found. I've found them still blooming in mid-October in the high country and well into the winter months along the coast. I photographed this pair along the upper Trinity River in early July and the inset at Mendocino on the coast in August.

COBWEBBY INDIAN PAINTBRUSH
Castilleja arachnoidea
Broomrape family

FIELD OWL'S-CLOVER
Castilleja campestris subsp. *campestris*
Broomrape family

CHICKWEED MONKEYFLOWER
Erythranthe alsinoides
Lopseed family

GIANT MONKEYFLOWER
Erythranthe guttata
Lopseed family

MUSK MONKEYFLOWER
Erythranthe moschata
Lopseed family

5 irregular petals forming a tube

This little lemony-yellow monkeyflower is easy to ID. The flowers have long corolla tubes with fused petals that flare away at the tips. Three of the petals are lined with deep reddish-brown stripes and are bearded in the throat. The shape of the flower, when looking directly into it, is round but each petal is notched at the tips with ruffled edges. The bright green leaves are more or less oval and can be slightly toothed at the margins. All the green parts of the plant are densely glandular and hairy making them sticky with a slightly musky smell. Plants I've found are10" - 12" in height but they can can be up to 24" tall. They favor damp meadowy areas and the edges of streams. I shot the large photograph above Adams Lake in the Trinity Alps in September and the small photograph illustrating the long corolla tube in the meadows along Canyon Creek in August.

PRIMROSE MONKEYFLOWER
Erythranthe primuloides
Lopseed family

5 irregular petals forming a tube

A singular deep red spot centered at the throat of each of the three lower petals of this monkeyflower make it a dead give-away in this area, but some have small red spots scattered in the throat of the flower as well. The petals are deeply notched at the tips and rounded. Look for the mats of oval or round leaves that can be either hairy or smooth. A single flower blooms on a long, thin, and dark pedicel from short tufted stems. They are rarely over 5" tall. Look for it growing in colonies in damp meadows, rocky seeps, or the edges of streams. It likes crevices in cliff faces where it gets lots of sun and moisture. I photographed this one at Stoddard Lake in the Trinity Alps in early September.

SHASTA LIMESTONE MONKEYFLOWER
Erythranthe taylorii
Lopseed family

5 irregular petals forming a tube

Climb around in the abundant limestone above Shasta Lake early in the season and you will likely be rewarded with this lemony-yellow monkeyflower. Small colonies bloom in cracks and crevices throughout the area tending towards sunny exposures. It is a small monkeyflower, most reaching only about 2" to 4" in height with deep reddish stems and gray-green fuzzy leaves which are toothed at the margins. The calyx tube is long with prominent red ridges and light red to pink between. Two bright red dots decorate the lower petal. Smaller red dots descend down into the flower tube as well. I photographed these above the McCloud arm of Shasta Lake in April. Now that extensive surveys have been done, this species has been found on substrates other than limestone but still geographically limited to the area around Shasta Lake. **Note:** it is often associated with small liverworts (see inset).

TILING'S MONKEYFLOWER, LARGER MOUNTAIN MONKEYFLOWER
Erythranthe tilingii
Lopseed family

5 irregular petals forming a tube

I've only run across this little monkeyflower in one place and that was above Mill Creek Lake in the Trinity Alps where it was growing in thick masses in wet, rocky clefts. In size, they are halfway between the chickweed and giant or common monkeyflowers. Their leaves are small, somewhat toothed, and a deeper green. The flowers are more orange-yellow but spotted with red on the lower petal. Unlike the giant monkeyflower, the flowers are solitary in the axils of the leaves. I photographed these in the rocks above Mill Creek Lake in the Trinity Alps in July (see small photograph).

MUSK MONKEYFLOWER
Erythranthe moschatus
Lopseed family

PRIMROSE MONKEYFLOWER
Erythranthe primuloides
Lopseed family

SHASTA LIMESTONE MONKEYFLOWER
Erythranthe taylorii
Lopseed family

TILING'S MONKEYFLOWER
Erythranthe tilingii
Lopseed family

DALMATIAN TOADFLAX

Linaria dalmatica subsp. *dalmatica* — 5 irregular petals forming a tube

COMMON TOADFLAX, BUTTER AND EGGS

Linaria vulgaris

Plantain family

Here's a noxious invasive plant that is fond of the open, sunny flats along the Trinity River and some of its southern tributaries. It's considered a weed, imported from its native haunts in Europe and Asia, but I find the flowers beautiful anyway. The buttery- yellow, long-spurred flowers look like little snapdragons and are arrayed along the upper half of a tall spike. **Note:** a second toadflax grows in the same neighborhood –*Linaria vulgaris* or common toadflax also known as butter 'n eggs. The difference between the two is is in the leaf shape.. Common toadflax has flowers arranged in a dense cluster at the tops of the fowering stem (see inset). The photograph of dalmation toadflax was taken along the Canyon Creek Road in mid-June while the inset of common toadflax was taken along the Trinity River near Douglas City in early July.

CHAPARRAL HONEYSUCKLE

Lonicera interrupta — 5 irregular petals forming a tube

Honeysuckle family

I had difficulties locating this beautiful honeysuckle but I had to track it down because Julie Kierstead wanted it included in this edition of our guide. I looked in all the places she indicated but was unsuccessful. Then, in June, James Lee with the Trinity River Restoration Project found a small population near Junction City. The deep buttery-yellow flowers bloom in tight bunches at the ends of long stalks that twine through the surrounding brush. Stamens and anthers are prominent, mostly because the petals fold back upon themselves Leaves are opposite but conjoined beneath the terminal flower cluster, forming a "disk" around the stalk. I photographed these near Junction City in June and Julie has photographed them near Redding.

DALMATIAN TOADFLAX
Linaria dalmatica subsp. *dalmatica*
Plantain family

COMMON TOADFLAX. BUTTER AND EGGS
Linaria vulgaris
Plantain family

CHAPARRAL HONEYSUCKLE
Lonicera interrupta
Honeysuckle family

HEARTLEAF ARNICA
Arnica cordifolia
Sunflower family

Flowers in composite heads

One of the distinguishing features of this arnica is the loose configuration of the ray flowers around the pale lemon-yellow flower heads. The ray flowers vary from oval to long and slender on the same head—and are notched at the tips. Leaves are opposite, heart-shaped, and toothed. They bloom in large colonies in shaded, damp sites at mid- to high-elevations. I shot this photograph near the summit of Scott Mountain in late June but one of the most beautiful displays is along the upper reaches of the Duck Lakes Trail in the Russians where they bloom in large colonies in the wet meadows.

MOCK LEOPARD BANE
Arnica dealbata
Sunflower family

Flowers in composite heads

I am reasonably certain that one of the first things you will notice about this little arnica is the color and texture of the leaves. In the population I photographed the leaves were mostly basal with a small number emerging along the flowering stalk. The shape was oval to egg shaped—tapered at the ends—and a striking blue-green. The texture was soft and slightly fuzzy. The bright yellow ray flowers were fairly large folding away from the central disc flowers in what is called a hemispheric shape. This particular population was quite large and covered a lot of ground near the East Weaver Lake Trailhead in the Trinity Alps. Plants ranged to about 8″ to 10″ tall. I photographed these in early July

RAYLESS ARNICA
Arnica discoidea
Sunflower family

Flowers in composite heads

This is a common arnica at lower elevations in these mountains. The flowering stems are fairly tall—upwards of 18″ in the population I found. Stems, leaves, and involucres are fuzzy. The deep yellowish-orange disc flowers are readily recognizable—blooming in compact heads that are mostly nodding at the ends of the flowering stalks. The population I found was growing in a fairly open but well shaded location beneath a mixed canopy of black oak and ponderosa pine. I shot this photographed on Baxter Ridge above Trinity Lake in June.

CLASPING ARNICA
Arnica lanceolata
Sunflower family

Flowers in composite heads

Several species of arnica are found in the area covered by this guide and are difficult to tell apart but the leaves and habitat make identification fairly easy. The sharply toothed leaves are lance-shaped, basal, and somewhat folded along the flowering stalks. For the most part the plants are strongly upright and found growing in rocky clefts near running water. The flowering stalks generally support one to three buttery-yellow composite flowers. The ray flowers are notched at the tips. I found and photographed these along Canyon Creek above Middle Falls and again at Upper Falls in late June.

HEARTLEAF ARNICA
Arnica cordifolia
Sunflower family

MOCK LEOPARD BANE
Arnica dealbata
Sunflower family

RAYLESS ARNICA
Arnica discoidea
Sunflower family

CLASPING ARNICA
Arnica lanceolata
Sunflower family

BROAD-LEAVED ARNICA, MOUNTAIN ARNICA

Arnica latifolia
Sunflower family

Flowers in composite heads

As opposed to heartleaf arnica, these flowers often bloom in loose clusters at the tops of their tall stems. Perhaps the best show of broad-leaved arnica, anywhere in this area, is along the last mile or so of the trail to Big and Little Duck Lakes. The shaded, moist meadows are loaded with them.These were photographed just below Little Duck Lake in the Russians in early August.

SEEP SPRING ARNICA, LONGLEAF ARNICA

Arnica longifolia
Sunflower family

Flowers in composite heads

Even on a good day amateur botanists, like myself, have a hard time identifyig the various arnicas found throughout this region. That being said, you can be reasonably sure that those beautiful golden-yellow flowers you've found clustered in large patches next to flowing streams and springs are, indeed, seep spring arnica. They are especially beautiful on the stream washed granite along the upper reaches of Stoney Creek (where I shot the large photograph) and in niches alongside the waterfalls along Canyon Creek (see small photograph). The paired leaves have lightly toothed margins, are long and narrow, and mostly fused along the flowering stems. Leaves and stems can be sticky to the touch. I photographed these in late July.

SHASTA COUNTY ARNICA

Arnica venosa
Sunflower family

Flowers in composite heads

This arnica has a strong upright habit, with dark green, lance-shaped leaves scattered in opposite pairs along the stems. The leaves are lightly toothed and somewhat folded along the mid-vein. The heads lack ray flowers and have mostly upright yellow disk flowers. There is the possibility that it could be mistaken for Rayless arnica (*A. discoidea)* but this one has smaller flowers and narrower, deeply-veined leaves. I shot these photographs along the McCloud Arm of Shasta Lake in June but have also photographed them along the shoreline of West Boulder Lake in the Trinity Alps. They are also found on Shasta Bally above Whiskeytown Lake and in the Squaw Creek watersheds as well.

BROAD-LEAVED ARNICA
MOUNTAIN ARNICA
Arnica latifolia
Sunflower family

SEEP SPRING ARNICA
Arnica longifolia
Sunflower family

SHASTA COUNTY ARNICA
Arnica venosa
Sunflower family

DELTOID-LEAF BALSAMROOT
Balsamorhiza deltoidea
SILKY BALSAMROOT
Balsamorhiza sericea
DELTOID-LEAF - SILKY BALSAMROOT HYBRID
Balsamorhiza sericea X deltoidea
Sunflower family

Flowers in composite heads

There is a place along the Upper Trinity River called Sunflower Flat which was probably named by miners coming into the area during the gold rush or settlers making their way up the old California Oregon Trail over Scott Mountain in the mid to late 1800s. I am convinced the name came from the balsamroot that blooms there in profusion. Of course, it would have attracted the attention of those early immigrants into the area. It is beautiful, bright, sunny and, when in full bloom, hard to ignore. Today, as you drive along Highway 3 in the upper reaches of the river above Coffee Creek in early spring, it can still gladden your day, much as it must have over 100 years ago—or thousands of years ago when Native Americans settled the area. The masses of beautiful, buttery-yellow flowers demand attention and are just three more reasons why botanists love this area so much.

Deltoid-leaf balsamroot (*B. deltoidea)* is native throughout California and is not confined to any particular soil type. Silky Balsamroot (*B. sericea*) on the other hand, is limited to the serpentine soils of the Klamath Mountains and is included in the California Native Plant Society list of Rare and Endangered Plants. I have no information about hybrid balsamroot (*sericea X deltoidea*), other than that it is found along with *deltoidea* and *sericea*, often in overlapping populations.

While the flowers of all three are pretty much the same, the leaves are quite different—which makes identification fairly easy. The deep green leaves of *B. deltoidea* vary in size but not so much in shape. Typically, they are large, wide at the base, and tapering to the tip—from which they gain their common name deltoid-leaf. The leaves of *sericea*, on the other hand, are gray-green, slightly fuzzy, pinnately compound, and deeply divided right down to the central stem. Leaves of the hybrid form - *sericea X deltoidea* - are a neat cross between its two contributors being deep green and pinnately veined and more moderately lobed.

DELTOID-LEAF BALSAMROOT
Balsamorhiza deltoidea
Sunflower family

SILKY BALSAMROOT
Balsamorhiza sericea
Sunflower family

HYBRID BALSAMROOT
Balsamorhiza sericea X deltoidea
Sunflower family

GRAY CUSHION PUSSYTOES, LOW PUSSYTOES
Antennaria dimorpha Flowers in composite heads
Sunflower family

It seems that every visit to Scott Mountain turns up something new and this unusual little pussytoe was no exception. I found a small population of four plants growing in a low, 3" - 4" tall, spreading mound covering an area roughly two by three feet. The spoon shaped leaves, sharply up-turned at the margins, were green but covered with long gray hairs giving them a decidedly blue-green look. This species is dioecious—meaning some plants are male and others female. These photographs are of the female plant bearing pistillate flowers. The flowers are dingy-brown fading a pale dark-yellow with prominent dark-brown pistils. I photographed these in early June.

WESTERN HAWKSBEARD
Crepis occidentalis subsp. *conjuncta* Flowers in composite heads
Sunflower family

This is another wildflower I've found in abundance on Scott Mountain. It grows in scattered locations on open, sandy to rocky sites on south and west facing slopes. The flower heads are typical of most hawksbeards—yellow with orange ligulate flowers. petals toothed at the tips with fairly prominent stamens, and anthers. What should really grab your attention are the leaves which are unique to the hawksbeards in this area. They are deeply cleft and wrinkled while being stiff to the touch and silvery gray-green. Leaf segments are somewhat cupped. They form a low basal mound from which the flowering stalks extend. Smaller leaves are found scattered along the stalks. Multiple flower heads bloom on naked stalks extending from the leaf axils. They often begin to bloom in mid-May and continue on through early summer. I shot this photograph in early July.

LARGEFLOWER HAWKSBEARD
Crepis occidentalis ssp. *pumila* Flowers in composite heads
Sunflower family

Here is another Scott Mountain denizen. Is it any wonder I love this place? The sheer variety and certain opportunities for discovery are enough to keep me coming back time after time. Through the years I have probably wandered past this little beauty repeatedly, thinking it was western hawksbeard. It finally dawned on me that the leaves were entirely different. These are dark-green, long, narrow, and pinnately lobed—which are tightly spaced. Flower heads of the two species are strikingly similar and, from a distance, difficult to differentiate. They prefer the same habitat—exposed, rocky slopes and flats and, on Scott Mountain, serpentine soils. I photographed these in late April on the west facing slopes of Scott Mountain.

NAKED STEM HAWKSBEARD
Crepis pleurocarpa Flowers in composite heads
Sunflower family

This hawksbeard is found throughout the area covered by this guide. It is common and widespread over several soil types and not too picky about habitat. The buttery-yellow ray flowers are narrow and notched at the tips. They bloom in a loose cluster along the upper part of the stalk. There are no disc flowers but there are prominent yellow styles, stamens, and anthers. The flowering stalks are, true to its name, naked—with long, narrow, lance-shaped leaves at the base. I shot these photographs in the wooded areas around Scott Mountain Campground in late June.

GRAY CUSHION PUSSYTOES, LOW PUSSYTOES
Antennaria dimorpha
Sunflower family

WESTERN HAWKSBEARD
Crepis occidentalis subsp. *conjuncta*
Sunflower family

LARGEFLOWER HAWKSBEARD
Crepis occidentalis ssp. *pumila*
Buttercup family

NAKED STEM HAWKSBEARD
Crepis pleurocarpa
Sunflower family

SPRING GOLD
Crocidiuim multicaule
Sunflower family

Flowers in composite heads

This little wildflower stood out like stars on a crisp dark night while driving along Highway 299 to the coast a few years ago. Not too far beyond Big Bar, on the Trinity River, I spotted these on the uphill side of the highway. It was the middle of February and the bright blooms caught me by surprise—seeming a month too early for such flowers to be front and center. Golden-yellow ray flowers are oval tapering to a rounded with a slightly notched tip. The knob of disk flower is a bit more orange. The flowers, in general, are about 2″ to 3″ in diameter. Leaves are bright green and oval, tapering to a pointed tip. The population I found consisted of about 2 dozen plants. These photographs were taken in Mid-February.

GREENE'S GOLDENBUSH
Ericameria greenei
Sunflower family

Flowers in composite heads

This member of the Sunflower family is common and widespread throughout the area covered by this guide. It displays lemon-yellow flower heads that appear to be structured haphazardly with long, narrowly oval, and pointed ray flowers. The disc flowers are small but the stamens and anthers are prominent. All in all, when blooming, the plant has a shaggy look. In the area where I took this photograph the plants formed low, mounded bunches up to 12″ or so in height and 24″ or so in diameter. It has an affinity for serpentine soils but is not necessarily limited to that substrate. I photographed this bunch near the summit of Mt. Ashland in southern Oregon where it was blooming on gravelly soils

FLEABANE DAISY, BLOOMER'S DAISY
Erigeron bloomeri var. *bloomeri*
Sunflower family

Flowers in composite heads

Rayless heads of bright yellow disk flowers bloom atop short reddish stems which are surrounded by masses of long, narrow leaves. The entire plant rarely exceeds 6″ to 8″ in height. It prefers open, rocky sites in full sun. I photographed this tiny daisy along the Pacific Crest Trail above Kangaroo Lake at the north end of the Trinity Divide in July. I've also seen it blooming atop the pass above Mumford Basin in the Trinity Alps. *Erigeron bloomeri* var. *nudatus* is also present in the book area and is ranked as rare.

ROCK DAISY
Erigeron petrophilus
Sunflower family

Flowers in composite heads

This tiny rayless daisy grows in open, rocky areas at mid- to high elevations. In passing, the flowers only vaguely resemble those of Bloomer's daisy but while this one has multiple flower heads per stem, *E. bloomeri* has only one per stem. Also, this one has lots of leaves along the stem while the leaves of *E. bloomeri* are always basal with no leaves on the stems. The flower heads are ldeep yellow, and bloom in small clusters at the tops of the stems. Both the leaves and stems are fuzzy. I shot this photograph on the ridge above East Weaver Lake in the Trinity Alps in early July.

SPRING GOLD
Crocidium multicaule
Sunflower family

GREENE'S GOLDENBUSH
Ericameria greenei
Sunflower family

FLEABANE DAISY, BLOOMER'S DAISY
Erigeron bloomeri var. *bloomeri*
Sunflower family

ROCK DAISY
Erigeron petrophilus
Sunflower family

COMMON WOOLY SUNFLOWER - two versions
Eriophyllum lanatum
Sunflower family

Flowers in composite heads

Several varieties of this species are found in the area covered by this guide–differing mostly in leaf color, hairiness, and degree of leaf lobing. The two pictured here show a common variety blooming from the floor of the Central Valley into the foothills of the surrounding mountains. The other I find quite frequently at mid- to high-elevations in the backcountry. The low elevation version is characterized by soft and fuzzy gray-green leaves sprouting all along the same fuzzy, gray-green stems. The high country version is characterized by slightly fuzzy dark-green leaves and nearly naked stems. The yellow blooms of both versions are identical. The low elevation version can be found blooming in large colonies or in scattered populations in places where they are exposed to dry and hot conditions (the cut banks of roads for example). These tend to bloom throughout the season, often coming on in early May and lasting well into August. I photographed these along Rush Creek Road near Lewiston in July. The high elevation version tends to bloom in small, scattered populations or individually at the edges of dry meadows near mixed-conifer stands. They begin their bloom at the end of July and continue on well into September. I photographed this one on Scott Mountain in early June.

BIGELOW'S SNEEZEWEED
Helenium bigelovii
Sunflower family

Flowers in composite heads

No matter where you go in the high country you are likely to find this flower in bloom somewhere along the way. It is mostly a meadow dweller but likes other wet areas as well. It tends to bloom in large colonies with dozens if not hundreds of plants. It can reach a couple of feet in height and the flower heads can be quite large. Probably the most distinguishing feature of this flower head is the large rounded knob of yellowish-orange disc flowers in the center which is surrounded by lighter, buttery- yellow ray flowers. I might add that it is one of the most persistent of the late bloomers in the high country. I've found beautiful flowers well into the late fall season–even into early November. This photograph was shot in the meadows around Stoddard Lake and the inset on Mt. Ashland along the Pacific Crest Trail.

CALIFORNIA HELIANTHELLA, CALIFORNIA LITTLE SUNFLOWER
Helianthella californica var. *shastensis*
Sunflower family

Flowers in composite heads

I find this wildflower blooming in the serpentine soils common along the east side of the Trinity Alps and northern end of the Trinity Divide, as well as in the decomposed granitic soils found on the south end of the Trinity Divide. The tall central flowering stalk is a dark reddish-green and holds a singular yellow-orange, compact head of ray and disc flowers. Leaves are lance shaped and strongly upright in a basal mass surrounding the flowering stalk. They tend to congregate in small colonies at low to mid-elevations where it is shaded and damp. I photographed these along the road to the Long Canyon Trailhead but look for them along Trinity Dam Boulevard near Lewiston and all along Highway 3 as it skirts Trinity Lake in May and June.

low elevation

mid to high elevation

COMMON WOOLY SUNFLOWER
Eriophyllum lanatum
Sunflower family

BIGELOW'S SNEEZEWEED
Helenium bigelovii
Sunflower family

CALIFORNIA HELIANTHELLA
Helianthella californica var. *shastensis*
Sunflower family

PRICKLY HAWKWEED, SHAGGY HAWKWEED
Hieracium horridum
Sunflower family

Flowers in composite heads

I found this tiny hawkweed blooming in a scattered population on open, exposed, and sandy soil between rocky outcrops on the ridge between Lower Sugar Lake and Little Duck Lake in the Russians. The flower heads are approximately 3/8″ in diameter with rays that are squared and deeply toothed at the tips. The stems and leaves are a deep green and covered with long bristly white hairs. I shot this photograph in early July.

DWARF ALPINE GOLD
Hulsea nana
sunflower family

Flowers in composite heads

This beautiful little flower can be a real show-stopper on the climb to the summit of Mt. Eddy. It was in full bloom all along the trail above Upper Deadfall lake when I made the trip on a warm day in the middle of July. There is absolutely no mistaking this for any other wildflower in the vicinity. The thick, pubescent leaves are long, narrow, ruffled at the edges and form dense gray-green basal mats on open, exposed serpentine and volcanic slopes. The golden yellow ray and disc flowers bloom in composite heads atop somewhat fleshy stems about 6″ or so in height.

COMMON MADIA
Madia elegans
Sunflower family

Flowers in composite heads

This is a common wildflower found from low to mid elevations throughout this area. The plants are tall and loosely branched with several flowers blooming along each stalk. The bright yellow ray flowers are deeply toothed with or without faint to deep burgundy markings at the base surrounding the deep orange disc flowers. It's interesting to me that there is, more often than not, a mix of flower coloration on each stalk. This photograph illustrates that nicely. This is one of those flowers that blooms early and stays late–often well into the fall season. I photographed these along the road to the Stoney Ridge Trailhead in early June but you will see them blooming in large colonies next to roads and trails everywhere in these mountains.

PRICKLY HAWKWEED
SHAGGY HAWKWEED
Hieracium horridum
Sunflower family

DWARF ALPINE GOLD
Hulsea nana
Sunflower family

COMMON MADIA
Madia elegans
Sunflower family

Deadfall Lakes Basin is a popular destination for day hikers and backpackers. The hike into the lakes is easy and the entire route is scenic with great views of the Divide country. Wildflower hunting is my main reason for making the trip but why do the hike if you don't bag Mt. Eddy while you're there? It is the highest peak in the Klamath Mountains and home to the remnants of one of the earliest lookouts in Northern California. Once you leave Upper Deadfall Lake and start to the peak the way is open and exposed with no water–so on a warm day remember your water bottles and don't forget your camera. Views from the summit are spectacular.

LITTLE TARWEED, MEAGER TARWEED, SMALL TARWEED
Madia exigua
Sunflower family

Flowers in composite heads

This diminutive wildflower is not showy at all and will probably go unnoticed by most people. It grows somewhat haphazardly and rarely gets any taller than 4." The tiny composite heads vary in size from 1/16" to maybe 1/8" in diameter. The ray flowers are lemon-yellow and deeply cleft at the outer margins. Disc flowers are deep brownish-orange. Leaves and stems are hairy. It favors fairy open areas with a light overstory of mixed conifer and hardwoods. I shot these photographs along Clear Creek below Clear Creek Campground in late May.

SLENDER TARWEED, GUMWEED
Madia gracilis
Sunflower family

Flowers in composite heads

Slender tarweed is just that–small, delicate, and usually less than 5" in height. The ray flowers are a pale lemon-yellow, deeply notched into threes with each notch rounded at the tip. The disc flowers are tubular and a bit darker with flared petal tips. A long, dark pistil extends from the center of each disc flower with a flared opening from which two yellow anthers protrude and bend back along the pistil. Stems and leaves are fuzzy and somewhat sticky to the touch. I photographed these along the road to the Swift Creek Trailhead in early June.

Deadfall Lakes Basin from Mt. Eddy on the Trinity Divide

LITTLE TARWEED, MEAGER TARWEED
Madia exigua
Sunflower family

SLENDER TARWEED, GUMWEED
Madia gracilis
Sunflower family

WAXY CONEFLOWER
Rudbeckia glaucescens
Flowers in composite heads
KLAMATH CONEFLOWER
Rudbeckia klamathensis
Sunflower family

These are the local sunflower family giants found blooming in damp meadows or beneath an open understory of mixed conifers. They tower over most of the other plants in the neighborhood some reaching heights of 3′. The flowers of both are similar and striking–with beautiful yellow ray flowers that drape away from prominent knobs of brownish-green disc flowers. Each stem is crowned with a solitary flower. The somewhat bluish-green leaves of *R. glaucescens* are basal and lance shaped with margins that are mostly entire or lightly toothed. Leaves of *R. klamathensis* are similar but a deeper green with margins that can be either smooth, wavy or coarsely toothed and/or lobed. I photographed the waxy coneflower in a meadow along Horse Creek in the Trinity Alps in mid-July (small photograph) and the Klamath coneflower along Swift Creek in mid-August.

CALIFORNIA BUTTERWEED, CALIFORNIA RAYLESS BUTTERWEED
RAYLESS RAGWORT
Senecio aronicoides
Flowers in composite heads
Sunflower family

This is a "stand-up" plant found in open woodlands and other fairly dry forested areas. It often reaches heights of 18″ or more with stout stems holding thick, flat topped clusters of buttery-yellow rayless flower heads. The dark green stems are slightly cobwebby giving them a silvery look. Leaves are mostly ovate or elliptic, basal, long, and tapered with wavy margins. It is common throughout the foothills of this region and really can't be confused with anything else. Around Whiskeytown Lake, where I photographed this one, they were closely associated with knobcone pine, whiteleaf manzanita, toyon, and California black oak.

WAXY CONEFLOWER
Rudbeckia glaucescens
Sunflower family

KLAMATH CONEFLOWER
Rudbeckia klamathensis
Sunflower family

CALIFORNIA BUTTERWEED
Senecio aronicoides
Sunflower family

LAMBSTONGUE GROUNDSEL, LAMBSTONGUE RAGWORT
Senecio integerrimus var. *major* — Flowers in composite heads
Sunflower family

This is one of those wildflowers that is almost impossible to miss. It blooms from low- to mid-elevations across these mountains–usually in forested areas. Buttery-yellow ray and orange disc flowers form dense clusters atop tall 1′ to 3′ stems. The stems, and the almost basal leaves, are all covered whitish cobwebby hairs. These photographs were taken near Lower Albert Lake in the Russians, along Highway 3 near Bowerman Ridge, and in Parker Meadows in the Trinity Alps–all in late spring and early summer. **Note**: In the first edition of this guide, we identified this as single stem butterweed; we have dropped that common name.

ARROWLEAF RAGWORT, ARROWHEAD BUTTERWEED
Senecio triangularis — Flowers in composite heads
Sunflower family

This *Senecio* resembles lambstoungue groundsel but, while that has hairy stems and leaves, the leaves of this one are not hairy and are definitely arrow shaped. The ray flowers are longer and more loosely arranged around the central cluster of orange disk flowers and the flowering heads are not as compact. The plants of both species are about the same height but grow in different habitats. *Senecio triangularis* prefers wet places while *Senecio integerrimus* prefers the drier understory of mixed conifer forests. Anyone in a hurry might easily confuse them. I shot this photograph along Swift Creek and the inset in an open wooded area below Sugarpine Lake in late August–both in the Trinity Alps.

CANADA GOLDENROD
Solidago elongata — Flowers in composite heads
Sunflower family

This goldenrod prefer open sunny wet to damp sites along the edges of streams and/or in meadows from low- to mid-elevations. Plants are 24″ or more in height. They are a welcome bright spot in any wildflower garden and definitely photogenic! I photographed these on the trail between Upper and Lower Wright Lakes in the Marbles in July and along the Stewart's Fork Trail near Nancy Creek, also in July.

LAMBSTONGUE GROUNDSEL
Senecio integerrimus var. *major*
Sunflower family

ARROWLEAF RAGWORT
Senecio triangularis
Sunflower family

CANADA GOLDENROD
Solidago elongata
Sunflower family

GOAT'S BEARD, YELLOW SALSIFY
Tragopogon dubius
Sunflower family

Flowers in composite heads

Found almost everywhere at lower elevations, it populates disturbed areas along roads, fields, and any other place it can get a hold. The plants are tall–up to 2′ and topped with golden yellow flower heads. Ray flowers are long, slender, and toothed at the tips. *Tragopogon* has ligulate heads--no disc flowers; what looks like disc flowers are just short 'ray' flowers (technically ligulate flowers). Sharply pointed nvolucre bracts radiate from beneath the flowers. While the flowers are striking the plants are probably best known for their large dandelion like heads of wispy seeds. I've photographed these along highways throughout these mountains, especially Highways 3 and 299. The bloom usually begins in late May and can last throughout June.

MULE'S EARS
Wyethia helenioides
Sunflower family

Flowers in composite heads

Mule's ears are a lower elevation early spring favorite. The large, prominently veined leaves are basal, stand, and are slightly hairy on the underside giving them a silvery look. Overall, they appear in shape and habit much like their namesake mule's ears, so the common name is a given. The often large and showy flower heads are striking–the ray flowers being a beautiful buttery-yellow and the disk flowers a deep orange. They can be found blooming in large populations around Whiskeytown and Shasta Lakes and in the lower foothills surrounding Redding. I photographed these in Jones Valley near Shasta Lake in mid-March.

NARROWLEAF WYETHIA, NARROWLEAF MULE'S EARS
Wyethia angustifolia
Sunflower family

Flowers in composite heads

This pretty member of the sunflower family can be found in scattered populations along Trinity Dam Boulevard (between Highway 299 and Highway 3), the Weaverville Basin, and along Highway 3 north of Weaverville. It is distinguished by the rather tall, narrow, prominently veined, deep green leaves. The buttery-yellow flower heads are a bit smaller than those of mule's ears but otherwise the two are quite similar. Its preferred habitat is open mixed hardwood and conifer forests. It blooms in early to mid-spring in areas where it is damp. I shot these photographs along Trinity Dam Boulevard near Lewiston in mid-May.

GOAT'S BEARD
Tragopogon dubius
Sunflower family

MULE'S EARS
Wyethia helenioides
Sunflower family

NARROWLEAF WYETHIA
Wyethia angustifolia
Sunflower family

ROCKY MOUNTAIN POND LILY, YELLOW POND-LILY, BRANDY-BOTTLE, SPATTERDOCK, YELLOW COW-LILY

Nuphar polysepala Many petals

Waterlily family

This is a beautiful aquatic perennial that almost everyone will recognize. It blooms in low- to mid-elevation mountain lakes and ponds throughout these mountains. Lilypad Lake in the Trinity Alps takes its name from this plant which, by mid-summer, has covered most of the lake with its large, heart-shaped leaves and gorgeous blooms. The flowers are cup shaped and buttery-yellow with small scale-like petals surrounded by thick, showy sepals. By late summer the flowers and leaf stems have died back leaving only submerged, withered stems and rootstocks behind. I photographed this one at Horseshoe Lake in the Trinity Alps in August. Years ago my daughter found pink pond lilies blooming in Fox Creek Lake.

WATER PLANTAIN BUTTERCUP

Ranunculus alismifolius Many petals

Buttercup family

Visitors to high country meadows in late spring and early summer will likely find this beautiful buttercup in full bloom. As the winter snow melts and turns the meadows into marshy wetlands this flower, along with marsh marigold (page 124), really comes into its own. It loves its feet in water and you will often find it completely submerged in spring run-off. The glossy, almost waxy looking, deep yellow flowers bloom on short stalks rising from dark green lance-shaped leaves. I photographed these in the marshy meadows of the Parker Divide in the Trinity Alps in early June. Another great place to see them is in meadows around Scott Mountain. **Note:** There are several varieties of *R. alismifolius* in the book area

ROCKY MOUNTAIN POND-LILY
BRANDY-BOTTLE
Nuphar polysepala
Waterleaf family

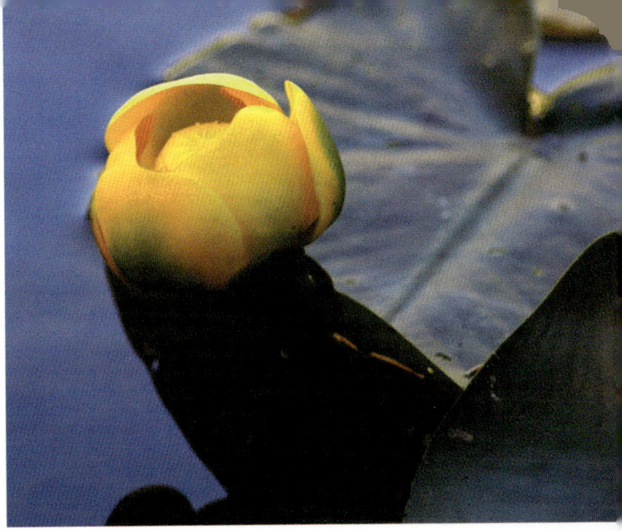

WATER PLANTAIN BUTTERCUP
Ranunculus alismifolius
Buttercup family

BUSH PENSTEMON
Keckiella lemmonii
Plantain family

5 irregular petals forming a tube

Talk about innocuous! The flowers of the bush penstemon are so small that hardly anyone ever notices them. Yet, when viewed up close, gorgeous flowers are found. The plants are hardly unassuming–some grow to over 4′ tall and frequently overhang low- to mid-elevation trails. They prefer partial shade and frequent the edges of mixed conifer and hardwood stands. When viewed from the top the flowers look like little brownish buttons but if you get down and take in the view from the bottom it's a whole different story. The hairy and sticky yellow and cream colored petals are striped with deep maroon and each flower has a prominent yellow staminode (sterile stamen) that is likewise hairy. Overall, the flowers are barely 1/4″ long and range in opposite pairs along the upper half of the stems. I photographed these along Slate Creek on the Trinity Divide but have found them blooming along the lower stretches of the trail to Duck Lakes in the Russians and along the Canyon Creek Trail in the Trinity Alps. You will also find them in the foothills of the Sacramento Valley where I've encountered them alongside trails near Keswick and Shasta Lakes.

ANTELOPE BRUSH, BITTER BRUSH
Purshia tridentata
Rose family

5 symmetrical petals - ovary superior

This is a stout shrub found in the relatively dry, open areas on the north end of the Trinity Divide and along the spine of the Scott Mountains. The shrubs I've found on Scott Mountain have ranged between 2′ and 6′ in height often forming low growing mats or scattered thickets. Stems are rigid and the twigs somewhat wooly. Leaves are gray-green, lightly hairy, alternate, wedge shaped, and three-toothed (hence "tridentata") at the ends. Like the twigs and leaves, the yellow to cream colored flowers are somewhat wooly. The petals are loosely arranged around prominent stamens which are topped with large orange anthers. I find them to have a very pleasant fragrance. I shot both of these photographs on Scott Mountain in mid-June when the bloom was in its prime.

SKUNK BRUSH, FRAGRANT SUMAC
Rhus aromatica
Sumac family

5 symmetrical petals - ovary superior

This shrub fakes out a lot of people who stumble into it and think they've wallowed in poison oak. Is it any wonder? Just look at the way the leaves are grouped! To top it all off, they belong to the same family. Skunk bush is innocuous however and will not give you that itchy rash that follows exposure to poison oak. The deciduous leaves of skunk bush are not oily or shiny and are more elongated but still in groups of three. The flowers are small, white to yellowish, and clustered at the ends of small, branched twigs. And while the berries of poison oak are small and white, the berries of skunk brush are pretty robust and always orange to red. It is common at lower elevations throughout the area. I shot the large photograph on Musser Hill near Weaverville in early July and the berries at Lowden Ranch near Lewiston in early October.

BUSH PENSTEMON
Keckiella lemmonii
Plantain family

ANTELOPE BRUSH
BITTER BRUSH
Purshia tridentata
Rose family

SKUNK BUSH, FRAGRANT SUMAC
Rhus aromatica
Sumac family

BIRCHLEAF MOUNTAIN MAHOGANY
Cercocarpus betuloides
Rose family

Alternate leaves and no obvious petals

I can almost guarantee that it will not be the tiny flowers that first attract your attention to this shrub. Instead, it will be the clusters of long-tailed, fuzzy, snow-white seeds (or achenes) that grab your eye. In this area birchleaf mountain mahogany rarely exceeds 15′ or so in height but it can grow to around 40′. The broadly oval leaves are prominently veined, persistent, and alternate with serrated margins in the upper half and smooth below. The trumpet shaped flowers are small and yellow with a reddish tinged base. There are no true petals on the flowers. It is a gnarly, rugged shrub that likes dry, sunny areas and is especially abundant on mid- to low-elevation mountain slopes. Look for it on hillsides along the Trinity River below Junction City, along the road to the North Fork of the Trinity River trailhead and along the road to the Hirz Mountain Lookout above the McCloud Arm of Shasta Lake.

CURLLEAF MOUNTAIN MAHOGANY
Cercocarpus ledifolius
Rose family

Alternate leaves & No obvious petals

This is one "gnarly" shrub with its twisted and deeply furrowed trunks, rough bark, and tangled branches. In this area it rarely exceeds 15′ in height and can be found at mid- to high-elevations throughout the Trinity Alps and Trinity Divide where it gets hammered by winter storms with fierce winds and deep snow. It gets its name from the narrow, deep green leaves which roll under from the margins. The small and trumpet shaped flowers vary from ¼″ to ½″ in diameter. They are white wth green centers and slightly reflexed with prominent stamens and anthers. That being said, it will probably not be the flowers that first attract your attention. More than likely it will be the beautiful clusters of feathery, snow-white seeds (or achenes) that, in late summer, turn the entire shrub a gauzy white. The large photograph was taken at the summit of Packer's Peak and the inset of the achenes above Mill Creek Lake–both in the Trinity Alps.

RUBBER RABBITBRUSH
Ericameria nauseosa
Sunflower family

Flowers in composite heads

Rabbitbrush decorates open, sunny mountain slopes and flats throughout the eastern Klamath Mountains and puts on quite a show when it blooms in mid-summer–turning entire areas a beautiful golden yellow. It forms thickets about 3′ tall with narrow, entire, gray-green leaves and dense heads of small golden-yellow flowers. It is a common shrub but look for it in areas that have been recently burned or that have a history of fire. I photographed these above Highway 299 near the French Gulch turnoff where entire mountainsides have burned through the years.

BIRCHLEAF MOUNTAIN MAHOGANY
Cercocarpus betuloides
Rose family

CURLLEAF MOUNTAIN MAHOGANY
Cercocarpus ledifolius
Rose family

RUBBER RABBITBRUSH
Ericameria nauseosa
Sunflower family

A cobra lily fen along the trail to Stoddard Lake in the Trinity Alps is also home to showy raillardella and bog asphodel

showy raillardella - *Raillardella pringlei*

LEOPARD LILY
Lilium pardalinum
Lily family

3 or 6 petals

This striking and graceful lily is a familiar sight along trail, streams, springs, and seeps throughout the high country. Falsely accused of being a "tiger" lily, I have to constantly remind people that tigers have stripes and leopards have spots. There are no stripes on this lily. This particular leopard, unlike its furry namesake, is not shy at all–sometimes blooming in huge colonies. There can be several flowers on stalks that reach over 6′ if conditions are right. Conditions are always right above Papoose Lake where huge numbers bloom in August. I shot the main photograph along the trail to Stoddard Lake in the Trinity Alps and the smaller one near lower Russian lakes in the Russian Wilderness. Recently I was surprised to find them in bloom along the eastern shoreline of Lewiston Lake. **Note:** There are three subspecies of *pardalinum* found in this area. 1) *L. pardalinum* ssp. *pardalinum* (common name - Leopard lily). 2) *L. pardalinum* ssp. *shastense* (common name - Shasta lily). 3) *L. pardalinum* ssp. *vollmeri* (common name - Vollmer's lily). Chances are good that you will run into any or all of them but differentiating between them is difficult.

WESTERN WALLFLOWER
Erysimum capitatum
Mustard family

4 petals

This wallflower is highly variable in color–ranging from pale yellow to deep orange. *Erysimum perenne* also occurs in the book area; it has yellow flowers. The plants can be up to 2′ tall though 12″ is more typical. It is found at lower elevations than its cousin the Sierra wallflower and is common along roadsides in dry, open, mixed conifer forests all the way from the coast to inland foothills. Flowers bloom in compact heads at the tops of the stalks, have four petals and strongly erect central stamens. I photographed these along the lower reaches of the Swift Creek Trail in late May. I have also photographed these along Gilman Road on the McCloud Arm of Shasta Lake.

COMMON FIDDLENECK, SMALL-FLOWERED FIDDLENECK
Amsinckia menziesii
Borage family

5 symmetrical petals forming a tube

The closer you get to this little wildflower the more beautiful and intriguing it is. Starting in late March, and lasting through early June, the orange and red spotted, trumpet-shaped flowers open as the tightly wound flowering stems begin to unfurl (they really do look like fiddle necks). The plants are tall (up to 18″) but the individual flowers are small–perhaps ¼″ in diameter when fully opened. The stems are covered with coarse reddish-brown hairs. From a distance the small flower clusters are not very showy but that changes when you get down at eye level. Look for them alongside trails, roads and highways in the area. They prefer open, exposed sites on flats and hillsides. I photographed these along Clikapudi Trail near Shasta lake but they are common throughout the foothills. The small inset is a portion of a watercolor painting I did years ago.

LEOPARD LILY
Lilium pardalinum
Lily family

WESTERN WALLFLOWER
Erysimum capitatum
Mustard family

COMMON FIDDLENECK
Amsinckia menziesii
Borage family

ORANGE HONEYSUCKLE
Lonicera ciliosa 5 irregular petals forming a tube
Honeysuckle family

Orange honeysuckle is fairly common in this area including in the Trinities. It is a wide-ranging vine often found clambering up trees and brush in protected and relatively damp sites. The brilliant reddish-orange blooms occur in bunches on short stalks arising from the centers of light green, conjoined oval leaves. In late summer and early fall brilliant clusters of red-cherry looking fruit develops making this vine just as showy as it is with the flowers. There is no mistaking this for any other wildflower in these mountains. I photographed these along Canyon Creek on the way to the Canyon Creek Trailhead in June where they are especially abundant. I've also found them blooming in various locations around Trinity Lake in mid-May.

SHOWY COLLOMIA, LARGE-FLOWERED COLLOMIA
Collomia grandiflora 5 symmetrical petals forming a tube
Phlox family

Not only is this Collomia beautiful but it can also be quite tall–ranging to 3′ in height. It prefers dry, open, or exposed brushy sites. Trumpet shaped flowers bloom in rounded, compact bunches at the ends of the stems. Delicate salmon, pale yellow, and deep orange flowers occur on the same flowering head. The petals flare sharply away from the tube encasing a snow-white stigma and beautiful light blue anthers. The blue anthers are striking and diagnostic for collomias. I shot these photographs along the trail between Lower and Upper Wright Lakes in the Marbles in late July. For a long time I thought it was restricted to areas in the high country but have since found it blooming in several sites around Trinity County including the Steiner Flat road above the Trinity River, Trinity Mountain Road above French Gulch, and Highway 3 along the Upper Trinity River.

KLAMATH MOUNTAIN CATCHFLY, SALMON-FLOWERED CATCHFLY
Silene salmonacea 5 symmetrical petals - ovary superior
Pink family

This is an amazingly beautiful wildflower, but because of their size, coloration, and scattered nature of their populations they are difficult to find. They don't hide, but rather bloom in plain sight–forming low mounds on exposed serpentine soils where they blend in wth their surroundings. There can be several 1 ½″ to 2″ salmon colored flowers blooming on each plant, each with five deeply cleft petals and prominent stamens. The stiff, basal leaves are a light olive-green and covered with soft silvery hairs. They are restricted to northern Trinity County and inhabit openings in Jeffrey pine/incense-cedar woodlands. I've found and photographed them blooming along the road to the Lake Eleanor and Poison Canyon Trailheads in late May and early June. They are especially fond of areas that have been recently logged and exposed to lots of sun.

ORANGE HONEYSUCKLE
Lonicera ciliosa
Honeysuckle family

SHOWY COLLOMIA
Collomia grandiflora
Phlox family

KLAMATH MOUNTAIN CATCHFLY
Silene salmonacea
Pink family

ORANGE AGOSERIS
Agoseris aurantiaca, var. *aurantiaca*
Sunflower family

Flowers in composite heads

The rusty-orange flowers of this species are a fairly common sight in dry, open areas at mid-elevations in the high country. I've never seen it in large colonies—and although you may find a lot of them in the same area, they do not seem to congregate. They are instantly recognizable, if not for the color, then for the long, rectangular and sharply toothed ray flowers. Each flowering stalk is crowned with a single flower head. I photographed this one on Mt. Ashland in southern Oregon but look for them along the trail to Doe Lake and in places like Steveale Meadows in the Trinity Alps. **Note**: this flower gets around. I have also photographed it high in the Grimmialp in Switzerland where it is called gold pippau.

SHOWY RAILLARDELLA
Raillardella pringlei
Sunflower family

Flowers in composite heads

This beautiful wildflower is endemic to the serpentine soils found in the northeastern Trinity Alps, the northern end of the Trinity Divide, and the Scott Mountains. I've never seen it bloom in abundance but have found it in isolated populations along the trail to Stoddard Lake, near Mill Creek Lake, and in Mumford Meadows—all in the Trinity Alps. A single, tall, almost gangly stalk, grows from a group of basal and mostly opposite, lanceolate leaves which are usually hidden in tall grass. The "single" flower blooming at the top of the stalk is made up of intense orange or yellowish-orange ray flowers and a central bunch of similarly colored disc flowers. The flower heads in this photograph were large—almost 3½" across and in their prime in early June. It's a real treat to find these but watch where you walk if you want to view them up close as the areas they occupy are usually boggy (or should I say "fenny" as they share the same habitat as cobra lilies and bog asphodel). I shot these photographs in a wet meadow near Gumboot Lake on the Trinity Divide.

Gumboot Lake: This is a popular lake for campers and day-trippers especially from the Mt. Shasta area. The meadows surrounding the lake and along the highway leading to it are loaded with wildflowers, including the showy raillardella. The Pacific Crest Trail passes above the lake on the south side and offers up magnificent vistas of Mt. Shasta and the surrounding country.

ORANGE AGOSERIS
Agoseris aurantiaca var. *aurantiaca*
Sunflower family

SHOWY RAILLARDELLA
Raillardella pringlei
Sunflower family

Gumboot Lake on the Trinity Divide

Applegate's paintbrush in the serpentine rocks surrounding Stonewall Pass in the Trinity Alps

Siskiyou Mountain owls clover and coyote mint in bloom along the Stoney Ridge Trail in the Trinity Alps

SIERRA ONION, DUSKY ONION
Allium campanulatum 3 or 6 petals
Onion family

Very often, the first sign that you are in the company of wild onions is the smell—but that isn't necessarily the case with this one. Instead, it's the size and beautiful color of the flowers. It's tall (up to 20" according to some sources) and showy. Each long stem carries a large, loose cluster of flowers numbering between 20 and 40 or more. The flowers in this area are mostly light magenta with cream centers ringed and spotted with a deeper magenta with a deeper magenta stripe running the central length of each petal. Stamens and anthers are light mauve and generally lay flat to the petals. Each plant has two or three flattened leaves, longer than the flowering stalks, which begin to turn brown and wither as the flowers open. I shot the main photograph under a canopy of mixed oak, fir, and pine on the ridgetop along Trinity Mountain Road near Smith Trail in late May and the inset on the Packer's Peak Trail in early June.

SCYTHELEAF ONION, SICKLE LEAF ONION, FALCATE ONION
Allium falcifolium 3 or 6 petals
Onion family

My first experience with this onion was on the open, exposed serpentine slopes of Scott Mountain several year ago. There were so many of them that those same slopes, seemingly barren in the summer, were a vibrant magenta at the height of the bloom. It is definitely a sight to see. The beautiful little upright vase-shaped light to dark purple flowers bloom in a thick rounded bunch atop a short 3" to 5" tall, thick, and flattened stem. Leaves are thick and basal—sometimes arching upward and away from the flowering stalk—usually lying flat on the ground. I shot the large photograph on Scott Mountain in mid-July on a slope below the Pacific Crest Trail before it crosses Highway 3. I found a nice population on the open serpentine slopes around East Boulder Lake in the Trinity Alps in mid-June.

DWARF ONION
Allium parvum 3 or 6 petals
Onion family

This is an early spring to late summer bloomer at mid- to high-elevations always on exposed, rocky soils. It seems not to be particular about the substrate and can be found on decomposed granitic as well as serpentinized soils. The plants are tiny, hardly ever growing more than 2" in height, with dense clusters of cream colored flowers veined with red, or white flowers veined with deep magenta. Often the only indication that you are in the area with them is the distinct onion smell released when you step on them! The main photograph depicts them at about 3X their actual size. Good places to find them are on the open gravelly slopes above Little Duck Lake where I photographed the cream and red version and along the road to the North Fork of Swift Creek Trailhead where I photographed the white/magenta version.

SIERRA ONION, DUSKY ONION
Allium campanulatum
Onion family

SCYTHE LEAF ONION
SICKLE LEAF ONION
Allium falcifolium
Onion family

DWARF ONION
Allium parvum
Onion family

SISKIYOU ONION
Allium siskiyouense
Onion family

3 or 6 petals

I first became aware of this striking onion the moment I stepped out of my truck at the East Weaver lake Trailhead, and it wasn't the flowers but the delicious smell of onions that attracted my attention. So intent was I on gathering my day pack and camera gear for my hike that I completely failed to notice the beautiful blooms all around my feet. The population there was immense and spread along the trail for several hundred yards. The soft lilac to deep reddish-purple flowers were arranged in concentric mounded heads at the tops of thick stalks. The fleshy leaves were long and slender and light to dark green in color. There are several species of onions found in these mountains, but these are among the showiest. I photographed these in early July.

PACIFIC MOUNTAIN ONION, WESTERN SWAMP ONION, TALL SWAMP ONION
Allium validum
Onion family

3 or 6 petals

This is the giant of the onion family in these mountains. I've found them up to 4' tall but usually around 2'. They like their feet in water preferring fens, wet meadows, seeps, and areas alongside creeks. The beautiful flowers fade from white near the base to a beautiful pale lavender near the tips and bloom in loose clusters at the tops of the stems. Stamens and styles extending well beyond the petals, giving the flowers a ragged look. Step on one, or accidentally crush one, and you will smell like onion for the rest of the day! I photographed these in a fen along the trail to Stoddard Lake in August.

CALYPSO ORCHID, FAIRY SLIPPER
Calypso bulbosa
Orchid family

3 or 6 petals

Calypsos are among my favorite early bloomers—often showing up in early May in moist, deeply shaded sites. A single, flower blooms atop a naked stalk that may reach 4" - 5" in height, above a single, broad basal leaf. The deep magenta sepals and lateral petals are alike—narrow and sitting like a crown above a hood of the same color. The hood arches over the top of the "moccasin" formed by the lower petal. The moccasin itself is whitish, hairy, and speckled with brownish-orange spots along the lower part of the lobe. Magenta spots descend into the throat. As is the case with all orchids they are very delicate and should not be disturbed. Enjoy them and take pictures but do not touch!

SISKIYOU ONION
Allium siskiyouense
Onion family

PACIFIC MOUNTAIN ONION
Allium validum
Onion family

CALYPSO ORCHID
Calypso bulbosa
Orchid family

FIRECRACKER FLOWER
Dichelostemma ida-maia
Brodiaea family

3 or 6 petals

The nodding, clustered, bright red blooms of this wildflower are unmistakable—easily identified and unforgettable. With a nod to their shape and color it's easy to understand how they got their common name. Locally they can grow to 3′ in height with the 1″ - 1½″ long tubular shaped flowers blooming at the ends of the tall, naked stems. Look for them beside lower elevation trails and roads throughout these mountains early in the spring. Trinity Mountain Road is a good bet for finding them in May as are Deadwood Road and Trinity Dam Boulevard near Lewiston.

SCARLET FRITILLARY
Fritillaria recurva
Lily family

3 or 6 petals

The brilliant reddish-orange bell shaped flowers of this fritillary are an eye popper. You will find them alongside roads and trails throughout the area at lower to mid-elevations. Flowers bloom in the axils of long, lance shaped leaves along the upper portion of a stalk that can reach 30″ in height. Its preferred habitat is shaded, moist sites, and it can frequently be found growing on the cut banks of roads. I photographed these alongside the road to the Eagle Creek Trailhead and Horse Flat Campgound in late May. Look for them as well as along the road to the Stewart's Fork trailhead, Rainier Road, or the Scorpion Creek Road to Bonanza King.

The Pacific Crest National Scenic Trail enters the Klamath Mountains east of Interstate 5 near Dunsmuir then courses through the Trinity Divide, the Trinity Alps, the Russian Wilderness, the Marble Mountains, and then over the Siskiyou Crest into Oregon. It is a magnificent trail system that invites people from all over the world to experience its rugged beauty. Thru hikers take pride in attempting the start to finish in record time—some doing 40+ miles per day. While that is an admirable feat, I find it a disconcerting that an obsession with mileage overshadows the time necessary to enjoy the little things that make these mountains so grand and beautiful. John Muir had it right when he said he didn't like the word "hike." He much preferred the term "saunter" because it suggested a slow and evaluative look at the landscape through which he wandered. How is it possible to observe the minutiae around us when we, in our obsession with speed, just race by. In my book it's the journey—not the goal that is important. Take to the PCT but remember that it provides the curious among us with a window into the magnificence of these mountains and the tiny and unobtrusive things that are the building blocks of that rugged beauty.

FIRECRACKER FLOWER
Dichelostemma ida-maia
Brodiaea family

SCARLET FRITILLARY
Fritillaria recurva
Lily family

Mt. Eddy from the Pacific Crest Trail above Gumbooth Lake on the Trinity Divide

OREGON ROCKCRESS
Arabis oregana 4 petals
Mustard family

Like Shasta rockcresses, this one also likes cracks and crevices in rocky outcrops. I've always found it in areas where shade is abundant although it does like some exposure to sunlight and it is almost always found in stream drainages near water. The plants are about 7" to 8" tall with rosettes of large, fuzzy basal leaves and smaller, lance-shaped leaves that develop alternately along the tall flowering stalk. The four petaled flowers bloom in the upper portion of the stalk and range in color from light to dark lilac. I photographed these on a rocky outcrop below Kangaroo Lake on the Trinity Divide. I've also found it in the Clear Creek Gorge south of Redding and along Clear Creek near the community of French Gulch.

SHASTA ROCKCRESS
Boechera breweri subsp. *shastaensis* 4 petals
Mustard family

The McCloud area around Shasta Lake is home to a variety of wildflowers that favor a limestone substrate and Shasta rockcress is one of them. This pretty little member of the mustard family can be found blooming as early as February in exposed, sun warmed clefts and crevices on west and south facing outcrops. Typically, the flowering stems are tall (8" to 16") with lance shaped leaves alternating to within an inch or two of the flower cluster on top. The basal leaves are hairy, more egg shaped, and rise out of a woody rootstalk. The leaves along the flowering stalk are less hairy. The pale lilac flowers have four petals that are reflexed with deep golden stamens recessed in the centers. I shot these photographs above the east side of the McCloud Arm of Shasta Lake in early February.

COMMON FIREWEED, TALL FIREWEED
Chamerion angustifolium subsp. *circumvagum* 4 petals
Evening-primrose family

This tall and showy fireweed is unmistakable. It can reach 5' or more with stalks crowned with large clusters of gorgeous, deep magenta-pink flowers that begin to show in July. It decorates meadows and the shorelines of lakes from the coast to the mountains during the entire summer. They derive their common name by the fact that they are often the first plants to colonize burned areas, like along the trail into Caribou Lakes pictured at right.. Good places to see them are in Bear Basin and Horseshoe Lake in the Trinity Alps and Russian Lake in the Russian Wilderness.

OREGON ROCKCRESS
Arabis oregana
Mustard family

SHASTA ROCKCRESS
Boechera breweri
subsp. *shastaensis*
Mustard family

COMMON FIREWEED
Chamerion angustifolium
subsp. *circumvagum*
Evening-primrose family

RED RIBBONS
Clarkia concinna subsp. *concinna* 4 petals
Evening-primrose family

Here is a beautiful flower that graces exposed, dry, and rocky roadsides and cutbanks in the lower elevations. It can grow to 2′ in height and is loosely branched with a multitude of flowers on each plant. The olive-green leaves are about 2″ to 3″ in length and oval to linear shaped. The blooms are striking—deep pink to red with some pale pink striping. There are four petals, each divided into three segments, with the middle segment wider at the ends than the other two—all ruffled at the tips. Good places to see them are along the French Creek and Canyon Creek Roads in May or early June. I photographed these along Canyon Creek Road in June.

LASSEN CLARKIA
Clarkia lassenensis 4 petals
Evening-primrose family

Occasionally, Mother Nature offers up a nifty surprise to those who least expect it. That happened to me while exploring forested areas along Canyon Creek Road in May when I ran into this pretty little clarkia. The stems are strongly erect, reddish-green, and covered with short, fine hairs. Flower petals are obovate forming a bowl, ruffled at the edges and a beautiful pinkish purple. The ovary is elongated below the flower, covered with fine hairs, and colored like the stem. The beautiful flowers really stood out against the dark green of the forest. It is also found at several sites in Trinity County and in the Russians along Sugar Creek.

WINECUP CLARKIA, PURPLE CLARKIA
Clarkia purpurea 4 petals
Evening-primrose family

This strikingly beautiful clarkia has diamond shaped petals that are deep purple at the throat surrounded by a whitish ring that fades to light purple. The petals are also folded longitudinally at the center and ruffled at the edges. The petals can sometimes be spotted. The style is snow white surrounded by creamy-yellow anthers. Overall, its habit is upright with a single flower atop each stalk. Leaves are more or less olive-green, linear and sharply pointed at the tips. These photographs were taken beneath a mixed canopy of oak, fir, and pine on a fairly dry bench above Canyon Creek in late May.

FAREWELL-TO-SPRING, DIAMOND CLARKIA
Clarkia rhomboidea 4 petals
Evening-primrose family

The name farewell-to-spring has a great ring to it in that it blooms near the end of spring and the beginning of summer (depending on elevation of course). It is a common clarkia in the foothills throughout this area. The plants can be tall and spindly—often reaching 2′—with small, lavender to wine-colored flowers blooming along the stems. The petals are narrow at the base and rhomboid shaped at the tips, white at the center, and spotted with deep magenta. Petal tips are ruffled. You can find them, if you're observant, along Rush Creek Road and along the Trinity River in May and early June. They are also common in the limestones above the McCloud Arm of Shasta Lake.

RED RIBBONS
Clarkia concinna subsp. *concinna*
Evening-primrose family

LASSEN CLARKIA
Clarkia lassenensis
Evening-primrose family

WINECUP CLARKIA, PURPLE CLARKIA
Clarkia purpurea
Evening-primrose family

FAREWELL-TO-SPRING, DIAMOND CLARKIA
Clarkia rhomboidea
Evening-primrose family

FOUR-SPOT, PURPLE CLARKIA, PURPLE GODETIA
Clarkia purpurea subsp. *quadrivulnera* 4 petals
Evening-primrose family

This little clarkia is found in damp meadows where it likes to hide in the tall grass. The stems are erect. Leaves are long, lance-shaped, and variably hairy. The four fan shaped lavender to wine-purple petals are small with a darker purple spot near each tip. Anthers are snow-white. A good place to look for them is along Rush Creek Road, Canyon Creek Road, and on Musser Hill near Weaverville.

PACIFIC BLEEDING HEART
Dicentra formosa 4 petals
Poppy family

This pretty and unique little wildflower has become a popular garden and landscaping plant. It is found all through the Pacific Northwest ut has been introduced to gardens world-wide. In the Klamath Mountains they begin their bloom in early May in the foothills—continuing well into August in the higher elevations. I have even found them blooming in late September just below Mirror Lake in the Trinity Alps. The rose pink flowers resemble fat, little, elongated hearts. They grow in tight clusters at the tops of the stems. The leaves are thin and deeply dissected, often appearing fern-like. Look for them in pockets along streams like the Upper Trinity River, Boulder Creek, Union Creek, and along the shoreline of West Boulder Lake in the Trinity Alps.

FEW-FLOWERED BLEEDING HEART
Dicentra pauciflora 4 petals
Poppy family

This diminutive bleeding heart loves serpentine. I've found vast carpets in places like the open slopes and rocky outcrops around Union Meadows and Union Lake. The feathery-leaved little plants grow from 2" to 5" in height. Each plant holds one to four nodding flowers ranging in color from near white to rose, pink, and sometimes deep purple. Each bloom has petals sharply reflexed at the ends. The prominent pistil is covered by the larger heart shaped petals, and extends beyond the flower as the seed pod matures. The tiny heart sits between the upright shoulders of the petals. They are early bloomers, often appearing as snow banks melt away beginning in mid-May at around 5000'. I photographed these along the Union Lake cut-off trail in the Trinity Alps in late May. They are common throughout these mountains.

FOUR SPOT, PURPLE CLARKIA
Clarkia purpurea
subsp. *quadrivulnera*
Evening-primrose family

PACIFIC BLEEDING HEART
Dicentra formosa
Poppy family

FEW-FLOWERED BLEEDING HEART
Dicentra pauciflora
Poppy family

CALIFORNIA FUCHSIA, HUMMINGBIRD TRUMPET
Epilobium canum subsp. *latifolium* 4 petals
Evening-primrose family

Brilliant displays of red trumpet shaped flowers with their extraordinarilly long pistils and stamens set this one apart from all other wildflowers in this area. The only other wildflower it could possibly be mistaken for is the redbush beardtongue shown on page 288 of this guide. The ball shaped anther should be the dead give-away if identification is in question. In the Trinities I've always found them growing on rocky outcrops and talus slopes at elevations of 6,000′ or more while redbush beard-tongue is found at much lower elevations. Leaves are deep green and both flowers and leaves are minutely hairy. A great place to see these is around Sapphire and Mirror Lakes in the Trinity Alps where they bloom in late summer and early fall. Barring a backpacking trip into the high country you can also find them blooming in the gardens of the Redding Arboretum where they are nicely tended. I photographed these just above Sapphire Lake in mid-September in the brushy slope shown in the small photograph.

DESERT WILLOWHERB, SLENDER ANNUAL FIREWEED
Epilobium minutum 4 petals
Evening-primrose family

I found and photographed this tiny epilobium along the trail to the summit of Packer's Peak in late June. There were scattered colonies of it blooming in the serpentine soils in shaded areas beginning at about 5,800′ then all the way along the trail to just below the summit at almost 8,000.′ As you can see, the teeny flowers (a little over 1/10″ in diameter) ranged from white to deep pink and occur mostly on pairs of stems originating in the leaf axils. Their preferred habitat is open and dry areas but they are also around vernal pools and disturbed areas. They frequently populate areas immediately after fire like its big cousin common fireweed.

SISKIYOU FIREWEED, ROCK FRINGE
Epilobium siskiyouense 4 petals
Evening-primrose family

You will find this beautiful little *Epilobium* growing in the crevices of serpentine rock outcrops and cliff faces above 5,500′. It rarely exceeds 10″ in height. Leaves are a bluish-green, somewhat fleshy, oval with pointed tips, and crowded along the stems. The leaves are usually coated with a light waxy bloom. One of the best places to see them is along the Fen Trail (where I photographed these and shown in the small photograph) above Kangaroo Lake from mid-July and on into early August. You will also find them in Bear Basin and on the slopes surrounding Seven-up Lake in the Trinity Alps. **Note:** Alice Jones, in her book, referred to this fireweed as "rock fringe" which is entirely fitting because it is often found decorating the edges and crevices of rocks.

CALIFORNIA FUCHSIA
Epilobium canum subsp. *latifolium*
Evening-primrose family

DESERT WILLOW HERB
Epilobium minutum
Evening-primrose family

SISKIYOU FIREWEED
Epilobium siskiyouense
Evening-primrose family

PACIFIC JEWELFLOWER
Streptanthus barbatus
Mustard family

4 petals

Populations of this little California endemic are limited to the serpentine soils of Trinity, Siskiyou, Shasta, and Tehama counties. Plants are found on dry, exposed, rocky areas and seem to like roadside cutbanks. The plants range in height from 8″ to around 12″ with stiff, woody stems. The stiff gray-green leaves are mostly round, clasp the stems tightly, and alternate in a tight pattern to within a few inches of the flowers—which also alternate up the stalk. The flowering buds are yellow- green and open vase-like in a deep purple—with slightly exposed anthers. I shot the left-hand photograph before the flowers opened along old Highway 3 near Eagle Creek Campground in early June. The other photographs of the mature flowers were taken alongside Highway 3 at the base of Scott Mountain in late June. You can find them blooming along the road from Highway 3 at Scott Mountain Summit to Big Carmen Lake as well.

FRINGE POD, HAIRY LACEPOD, SAND FRINGEPOD
Thysanocarpus curvipes
Mustard family

4 petals

This member of the mustard family is one of those that, until it goes to seed, is fairly unassuming. It is a leggy plant often growing from 15″ to 24″. The tiny white flowers bloom in an open raceme followed by seed pods that are as beautiful as they are interesting. These pods are what most people will remember about this plant. The pods form a translucent elongated oval with a central yellowish-green pod from which short green ribs radiate outward—quickly turning magenta to the edges. The interstices between the ribs form small openings near the central pod but are closed and white or light pink as they near the edge. They are showy and really put on a beautiful display when backlit by the sun. I photographed these along the Trinity River near Junction City in late June.

RIBBED FRINGE POD, SHOWY FRINGEPOD, SHOWY FRINGE POD
Thysanocarpus radians
Mustard family

4 petals

This leggy wildflower is common in open areas throughout the valleys and foothills covered by this guide. The tiny greenish-white flowers are probably not what will initially attract your attention. More than likely it will be the pretty little seed pods from which it gets its common names. The pods hang down from the ends of alternating reddish stems, are round and translucent, with green ribs radiating from a central yellowish-green pod outward to a thickened, deep magenta rib. The interstices between the green ribs and the outer magenta rib are mostly white. They are showy, especially when backlit by the sun. I took this photograph at Lowden Ranch along the Trinity River in early July.

PACIFIC JEWELFLOWER
Streptanthus barbatus
Mustard family

FRINGE POD, HAIRY LACEPOD
Thysanocarpus curvipes
Mustard family

FRINGE POD
RIBBED FRINGE POD
Thysanocarpus radians
Mustard family

RED LARKSPUR, CANYON LARKSPUR
Delphinium nudicaule
Buttercup family

5 irregular petals

This is a wildflower that frequents the sides of roads in the foothills and low elevation backcountry trails throughout the area. It especially likes damp, shaded hillsides and mossy rock outcrops. The flowers are ¾" to 1" long, deep orangish-red, with a long spur at the back. The petal-like sepals flare at the tips. They bloom in large numbers along the flowering stalk. Good places to see these are along Rainier Road above Trinity Lake, in the limestone formations along Gilman Road just above Shasta Lake, and along the road to the Stewart's Fork Trailhead in May—which is where I photographed these.

EVERLASTING PEA, SWEET PEA
Lathyrus latifolius
Pea family

5 irregular petals

I think the everlasting sweet pea fairy works overtime along Highway 3 and Trinity Dam Boulevard. Drive those highways any time during the blooming season and the beautiful flowers are seen non-stop. If ever there was a universal favorite among flower lovers this has to be it—though it's not a favorite of the weed crews who try to get rid of it on public lands. I honestly can't think of anyone who doesn't know this one. This European native is grown in gardens and has found its way to myriad roadsides, stream banks and forest edges at lower elevations all through these mountains. True to its name, everlasting pea blooms from early spring well into the fall all the way to the coast. Flower color varies widely, but I think that has more to do with the original colors of garden escapes through the years, and soil types. Look carefully at photographs of homes and cabins dating from the 1800s in Trinity County and you will almost always see them twining through garden fences.

SIERRA PEA
Lathyrus nevadensis var. *nevadensis*
Pea family

5 irregular petals

This little pea grows in large, scattered populations all along trails and roads at low to mid-elevations throughout the area. The blossoms are a delicate pinkish-magenta and the leaves are pinnately compound. Stems are jointed in a slight zig-zag shape between the leaves. They grow, or sprawl, low to the ground in areas with some tree cover but with adequate exposure to the sun—where they twine through downed branches and brush. Mumford Meadows in the Trinity Alps is a good place to see them as are the roadsides leading to the Granite Peak Trailhead along Buck Ridge (where I photographed these). You can also find them along Highway 3 between Covington Mill and Trinity Center.

RED LARKSPUR
CANYON LARKSPUR
Delphinium nudicaule
Buttercup family

EVERLASTING PEA, SWEET PEA
Lathyrus latifolius
Pea family

SIERRA PEA
Lathyrus nevadensis
var. *nevadensis*
Pea family

CHICK LUPINE
Lupinus microcarpus
Pea family

5 irregular petals

I originally found this beautiful lupine near Natural Bridge south of Hayfork and was surprised when it popped up along the road to the Canyon Creek trailhead. The tiers of deep pink flowers are arranged along tall, reddish-brown, 12′ to 18″ stalk. The leaves are also tinged with red at the margins. This is the only pink lupine in this area and, therefor difficult to mistake for anything else. It is widespread throughout California with only a few documented populations in Trinity and Siskiyou Counties. I photographed this one in early May along Canyon Creek. Possibly the most beautiful display of chick lupine I've ever seen was along the Ditch Grade between Platina and Highway 36 in late May and early April.

SHASTA CLOVER
Trifolium productum
Pea family

5 irregular petals

This a distinctive clover and almost impossible to confuse with any other. It displays descending, tubular shaped, white to deep purple flowers that bloom in compact heads atop tall, naked stalks. Its prefers damp, open locations at higher elevations. The flowering heads can be up to 1½″ long. The leaves are long and oval with toothed margins. I've found it throughout the Trinities, Russians, and Marbles. I shot the main photograph in the meadow above Upper Wright Lake in the Marbles and the small one in the wet meadow at the back of Big Duck Lake in the Russians.

SIERRA MILKWORT
Polygala cornuta
Milkwort family

5 irregular petals

The closer you get to this tiny wildflower the more beautiful it is. The white and magenta-purple flowers—each with two wing-like sepals—are gathered in loose clusters at the ends of short stems. Stamens and pistils are hidden by closely folded petals. The oval leaves are dark green. I've found it in various lower elevations throughout the Trinity Alps. I photographed these at the East Fork of Stewart's Fork Trailhead in July. I have also found it blooming alongside trails above Keswick Lake in the Sacramento River Canyon. It is common and widespread.

CHICK LUPINE
Lupinus microcarpus
Pea family

SHASTA CLOVER
Trifolium productum
Pea family

SIERRA MILKWORT
Polygala cornuta
Milkwort family

RED COLUMBINE, CRIMSON COLUMBINE, SITKA COLUMBINE

Aquilegia formosa

5 symmetrical petals - ovary superior

Buttercup family

This wildflower doesn't really need an accompanying description as it looks like no other found in this area. It is common from the low elevation foothills around Shasta Lake to mid-elevations in the mountains. They favor shaded, damp sites that, in the foothills, mean north facing slopes. At higher elevations they like wet, shaded, meadow areas. A great place to see them in the high country is along the trail to Big and Little Duck Lakes in the Russians where they grow in profusion along with dozens of other wildflowers that seek out similar habitats. I photographed this one along the Waters Gulch Trail above Shasta Lake in May.

RED MAIDS

Calandrinia menziesii

5 symmetrical petals - ovary superior

Miner's lettuce family

This brilliantly colored little wildflower likes open and sunny places where it blooms in large numbers. The deep rose colored flowers are about the size of your little fingernail and sit singly atop a short stem bracketed by small, pointed leaves. The basal leaves are stout, fleshy, and club shaped becoming more oval at the tips. All of the leaves are ciliate, or hairy, at the edges. It occurs commonly at lower elevations throughout the area and can be found alongside trails and roads. I photographed these alongside Clikapudi trail near Shasta Lake in late May but have also found them along Highway 299 along Trinity Dam Boulevard near Lewiston, in Lorenz Gulch on the Trinity River and along Canyon Creek Road.

PIPSISSEWA, PRINCE'S PINE

Chimaphila umbellata

5 symmetrical petals - ovary superior

Heath family

Pipsissewa is another of those outstanding "belly flowers." It hardly ever grows over 8" tall and is usually found in large populations. The plant is evergreen with stiff stems and deep green, oval leaves that are stiff and toothed. Flowers bloom in clusters and are mostly deep pink fading to near white at the tips of the petals. The petals arch away and form small "hoods" over a prominent green stigma surrounded by a crown of ten stamens. It prefers deep, shaded conifer woods and is quite common throughout these mountains, mostly at lower to mid-elevations. I photographed these above Sugarpine Lake in late September.

HAIRY ROCKROSE

Cistus incanus

5 symmetrical petals - ovary superior

Rock-rose family

This is a naturalized, non-native wildflower found in this area. I was most likely intentionally planted by the California Department of Transportation in a highway beautification project. The numerous large, showy flowers have a texture like crumpled tissue paper, are vibrant purple, with deep orange pistils and anthers. When in bloom can be seen from quite a distance. The shrubby plants grow to about 3' and have leaves and stems that are grayish-green. They thrive in disturbed areas and can be especially invigorated by wildfire. I photographed these along Packer's Bay Road above Shasta Lake but you will find a very nice display along Highway 299 W on Oregon Mountain in mid-June.

RED COLUMBINE, SITKA COLUMBINE
Aquilegia formosa
Buttercup family

RED MAIDS
Calandrinia menziesii
Miner's lettuce family

PIPSISSEWA, PRINCE'S PINE
Chimaphila umbellata
Heath family

HAIRY ROCKROSE
Cistus incanus
Rock-rose family

PINK GRASS, WINDMILL PINK, WILDING PINK

Petrorhagia dubia 5 symmetrical petals - ovary inferior

Pink family

Small but striking! That's how I would describe this naturalized member of the pink family. It can be found blooming in colonies of scattered individuals on sunny flats in and around rocky formations. The flowering stalks are tall—some reaching upwards of 18"—with narrow leaves that wrap around the stems like a sheath. The bulbous and elongated involucre sheaths produce a number of individual flowers that bloom successively, one at a time poking out from the sheath. Each pink petal is notched and striped with deep red near the white centers. I shot these photographs on a floodplain just above the Trinity River below Douglas City but sightings have been recorded between Lewiston and Big Bar.

WHITE MARGINED KNOTWEED. POLYGALA KNOTWEED

Polygonum polygaloides 5 symmetrical petals ovary superior

Buckwheat family

My son and I found this little member of the buckwheat family on a hike into Shimmy Lake in the Trinity Alps in late September. From a distance it appeared as an ephemeral purplish cloud floating above the sunbaked soil. Upon closer inspection— which always means crawling around in the dirt—the cloud became a beautiful masses of distinct white and purplish-pink tube-shaped flowers. The dark green eaves are tinged with red and scattered along the flowering stems. It appears that the weight of the flowers caused the flowering stalks to lie almost flat to the ground with a slight upturn at the tips. I've also photographed it on flats along the Salmon River above Big Flat.

HARSH CHECKERMALLOW, HARSH CHECKERBLOOM

Sidalcea asprella 5 symmetrical petals - ovary superior

Mallow family

My thesaurus equates "harsh" to bright, dazzling, garish, lurid, and gaudy. So, I suppose whoever named this wildflower had dazzling or bright in mind. This species prefers wet or damp meadows or open hillsides. The deep rose flowers bloom on a tall stalk and are found into October in protected areas. The striated petals are deeply notched at the tips and surround a pink pistil—topped with a unique encircling crown of white anthers Look for it all over the mid-elevations. I photographed these in Long Canyon Meadows in the Trinity Alps in August.

OREGON CHECKERMALLOW, OREGON CHECKERBLOOM, SPICATE CHECKERMALLOW

Sidalcea oregana 5 symmetrical petals - ovary superior

Mallow family

You can find this beautiful mallow blooming all summer long in the high country—even into October in protected areas. It likes wet places and can be found wherever there is abundant moisture. It grows to about 3' tall, usually in large colonies. The rose colored, bowl shaped flowers bloom in thick bunches at the tops of the stalks. The flower petals are striated with a deeper rose. I photographed these on Mt. Ashland in southern Oregon but look for them in Bowerman or Long Canyon Meadows in the Trinity Alps or along the Paynes Lake trail in the Russians.

PINK GRASS, WINDMILL PINK, WILDING PINK
Petrorhagia dubia
Pink family

WHITE MARGINED KNOTWEED
Polygonum polygaloides
Buckwheat family

HARSH CHECKERMALLOW
Sidalcea asprella
Mallow family

OREGON CHECKERMALLOW
Sidalcea oregana
Mallow family

INDIAN PINK, CALIFORNIA PINK
Silene laciniata subsp. *californica*
Pink family

5 symmetrical petals - ovary superior

These beautiful wildflowers are early bloomers in the foothills. Look carefully and you will find them decorating cut-banks along Highway 299, Highway 3, Trinity Mountain, and Rush Creek Roads. The brilliant red flowers cannot be mistaken for any other in the area. The similar red *Silene serpentinicola* approaches the book area from the west. They are about 1" across, with five deeply cleft petals and prominent stamens rising straight out of the center. The plants are loosely branched with long stems but spread laterally along the ground to 7" or 8" tall. I phot graphed these along Highway 299 near Buckhorn Summit in May.

FENDLER'S MEADOW-RUE
Thalictrum fendleri
Buttercup family

5 symmetrical petals - ovary superior

This unusual wildflower is one that a lot of hikers will never see, and not because it hides. It grows and blooms in plain sight in wet, shaded areas where other plant seeking the same habitat grow in profusion. In the mass of vegetation this one doesn't particularly stand out. It is dioecious, with one plant having female flowers and another male flowers—each very distinctive. The flowers are green and cup shaped. Male flowers have distinctive dangling clusters of stamens topped with long anthers. If you find one, you will also find the other because they often grow entangled. Overall, the plants reach upwards of 7' in height, and are loosely branched. The leaves look a lot like columbine. I photographed these in the wet thickets below Middle Bowerman Meadows in the Trinity Alps on the 4th of July.

CALIFORNIA CENTAURY, CHARMING CENTAURY
Zeltnera venusta
Gentian family

5 symmetrical petals ovary superior

This little beauty blooms early in the spring at lower elevations in the surrounding foothills. The flowers are deep reddish-purple with lance shaped petals that flare away from the center revealing a cluster of bright yellow spirally twisted anthers. Flowers bloom in a loose bunch at the tops of stems that can grow to about 7" or so in height. Look for them in open, grassy, vernally wet areas along Rush Creek Road and above Lewiston Lake (where I shot these photographs).

ENGELMANN'S DESERT PARSLEY, ENGELMANN'S LOMATIUM
Lomatium engelmannii
Carrot family

5 symmetrical petals - ovary inferior

The beautiful bunches of lilac colored flowers are the give-away for this small, ground hugging *Lomatium*. The leaves are long, deeply dissected, and along gray-green in color. Stems range from gray-green to deep red or greenish-red. This wildflower is so small and unobtrusive that most people will probably walk right over it without notice. It appears to favor serpentine soils in mostly open areas. I shot this photograph in early April along Eagle Creek near Highway 3 and have also photographed it on Scott Mountain in late May.

INDIAN PINK
Silene laciniata subsp. *californica*
Pink family

FENDLER'S MEADOW-RUE
Thalictrum fendleri
Buttercup family

CALIFORNIA CENTAURY, CHARMING CENTAURY
Zeltnera venusta
Gentian family

ENGELMANN'S DESERT PARSLEY
Lomatium engelmannii
Carrot family

PURPLE SANICLE
Sanicula bipinnatifida — 5 symmetrical petals - ovary inferior
Carrot family

This inconspicuous little sanicle could easily go un-noticed except for the leaves which are large and light-yellowish to olive green. They are deeply cleft and toothed growing alternately along a central light reddish to gray-green stem. The tiny 3/8″ to 1/2″ purple (and sometimes yellow) flowers grow in several ball-shaped heads atop a long, thin stem. They occur in large populations and seem not to be picky about soil type. I photographed the purple version along Trinity Mountain Road on the Trinity Divide in late April and the yellow version along the road to the Lake Eleanor Trailhead in the Trinity Alps in late May. **Note:** Is it purple or yellow? And, if it's yellow what is it doing in the red to purple section of this guide? It's both. Predominantly it is purple but with so many showing up yellow what's a classification nerd to do? Fall back and punt. Put it where the name indicates and let the yellow chips fall where they may.

GLAUCOUS TAUSCHIA, GLAUCOUS UMBRELLAWORT
Tauschia glauca — 5 symmetrical petals - ovary inferior
Carrot family

This wildflower is distinguished by the tall reddish-green stems which are topped with radiating clusters of tiny deep rose-purple flowers. Styles and stamens extend well beyond the petals. The leaves are a pale muted green, deeply cleft and toothed along the margins. This is a common wildflower found on open, grassy or rocky areas in serpentine soils at lower elevations. I shot these photographs along the road to the Lake Eleanor Trailhead in June.

FRAGRANT FRINGECUP, BIGFLOWER TELLIMIA, FRINGE CUPS
Tellima grandiflora — 5 symmetrical petals - ovary inferior
Saxifrage family

Botanical manuals indicate that the flowers of this plant are usually white but that there are red and pink variations. I've never seen one with white flowers in this area, only pink but I will keep looking! I've never detected any fragrance eminating from this wildflower. The fragrance molecules must be microscopic or my nose must be jaded. This is one wildflower you have to get up close and personal with to appreciate its delicate beauty. It grows in damp, shaded places and because the flowers are so tiny, it all blends in with the background foliage. The flowers are cup shaped with reflexed, deeply cleft, sharply pointed petals. They are scattered along stalks that in some places are 12″ to 16″ tall. Leaves, which cover the lower third of the stalk, are large, palmately lobed, and veined and almost heart shaped. These photographs were taken along Stewart's Fork Trail, VanNess Creek above Trinty Lake and Fender's Ferry Road above Lake Shasta, all in mid-May.

PURPLE SANICLE
Sanicula bipinnatifida
Carrot family

GLAUCOUS TAUSCHIA
GLAUCOUS UMBRELLAWORT
Tauschia glauca
Carrot family

FRAGRANT FRINGECUP
BIGFLOWER TELLIMIA
Tellima grandiflora
Saxifrage family

WOODLAND STARS

Lithophragma (several species) 5 symmetrical petals - ovary superior

Saxifrage family

The star-like, deeply divided, pale pink or creamy white flowers bloom on simple, tall stems. Each of the five petals are lobed. Several species of woodland stars have been noted in the book's area but it's a challenge to distinguish one from from the other. Here we feature pink woodland star (*L. parviflorum*), hillstar (*L. bolanderi*) and Siskiyou Mountain woodland star (*L. campanulatum*). These are found at mid-elevations frequenting the sides of backcountry trails. I shot these photographs on the Stewart's Fork, Canyon Creek and Swift Creek trails mostly in early to mid-June.

ROSEROOT STONECROP, PACIFIC ROSEROOT, KINGSCROWN, LEDGE STONECROP

Rhodiola integrifolia subsp. *integrifolia* 5 symmetricalpetals - ovary superior

Stonecrop family

The only place I've ever seen this is at Big Duck Lake in the Russian Wilderness where it grows in little pocket meadows scattered throughout rocky areas at the base of the cliffs towards the back of the lake. It is a small succulent with light green leaves that grow in whorls up the flowering stem. The flowers are beautiful, deep magenta, red, and yellow and bloom in a tight, flattened cluster at the tops of the stems. This is an absolutely gorgeous stonecrop and possibly the most beautiful of all in the Klamath Mountains!

WOODLAND STARS
Lithophragma (several species)
Saxifrage family

ROSEROOT STONECROP, PACIFIC ROSEROOT, KINGSCROWN, LEDGE STONECROP
Rhodiola integrifolia subsp. *integrifolia*
Stonecrop family

COBRA LILY, COBRA PLANT, CALIFORNIA PITCHER PLANT

Darlingtonia californica 5 irregular petals - ovary superior
Pitcher plant family

Here's a little meat eater for you! One of those vicious little plants that attracts bugs, invites them into its inner most workings, kills, and then eats them. Too bad for the bugs! The modified leaf that forms the plant resembles a hollow tube—narrow at the bottom where it arises from the rootstock, widening as it ascends, finally folding over at the top in a shape that resembles a rearing cobra—replete with a "leafy" appendage flaring away from the little opening at the bottom of the head that one could construe as "fangs." The plant varies in color from a beautiful light green to a mottled yellow-green with numerous maroon spots as they age. They range in size to 18" tall and are often found in huge colonies.

The trap, or opening, by which insects are enticed to enter the plant, is hidden beneath the downward pointing "head" and behind the "fangs." Once the insects are inside, numerous translucent windows along the leaves allow light in and confuse them (much like the windows in our houses attract flies). Also inside, the stems are stiff, downward pointing hairs, and a secreted lubricant—which further prevent escape. Pooled at the bottom of the stem is an enzyme laden liquid that, once encountered by the hapless insect, begins the digestion process. This provides nutrients to the plant not available from the serpentine soils on which they grow.

In late spring and early summer, large, deep red, inverted, vase-shaped flowers nod gracefully at the ends of tall red-green stems and are surrounded by pale yellow-green sepals. Cobra lilies are a "flagship" species throughout the serpentine country described in this guide, growing in huge colonies in fens and hillside seeps. Good places to see these carnivores at work are along the Swift Creek Trail in the Trinity Alps, the Parks Creek Road on the Trinity Divide, and the Fen Trail above Kangaroo Lake in the Scott Mountains. My son, Chris shot this gorgeous photograph of cobra lilies at Lake Eleanor in late September. I photographed the flower in early June near Bullard's Meadow on Union Creek and the seed pod near Eagle Creek in November. I shot the photograph of the fen along the Swift Creek Trail in July where there were also leopard lilies, Klamath coneflower, white hastingsia, and sneezeweed.

COBRA LILY, COBRA PLANT, CALIFORNIA PITCHER PLANT
Darlingtonia californica
Pitcher plant family

ONE-SEEDED PUSSYPAW
Calyptridium monospermum — 5 irregular petals - ovary superior
Miner's lettuce family

This pussypaw has one basal rosette of deep green, elongated, spoon shaped leaves with smaller leaves present along the flowering stem. There are two or more flowering stems per rosette. Flowers bloom in dense rose colored clusters at the ends of the stems which hug the ground at night—rising a few inches off the ground in full daylight (probably to avoid the heat). Typically the flowering heads are no more than 2" across. Look for them throughout the summer wherever you find exposed sandy flats. I photographed these on Mt. Ashland in July but they are common throughout the area. An especially nice display is along the Stewart's Fork Trail between Morris Meadows and its junction with the Suicide Trail.

PUSSYPAW, UMBELLED PUSSYPAW
Calyptridium umbellatum — 5 irregular petals - ovary superior
Miner's lettuce family

This pussypaw typically has two or more clustered rosettes of deep green, elongated, spoon shaped leaves—with no leaves on the flowering stem. There is one flowering stem per rosette and the flowers are typically white, tinged pink with purple anthers. Occassionally you will find white versions with deep red anthers. The plants rarely exceed 5" in height. Flower stalks are long and each is topped with a round ball of compact flowers about 2" in diameter. During the cool hours the stalks lie almost flat to the ground but as the day heats up the stalks rise until some of them are vertical. They prefer dry, sandy soils and can be found throughout the area from subalpine to low elevation locations. I shot this photograph near Middle Falls along Canyon Creek where they are abundant on the sandy flats between granite outcrops. I photographed the inset of the white version on the Siskiyou Summit in an area where this was the predominant color.

NETTLELEAF HORSE MINT, GIANT HYSSOP
Agastache urticifolia — 5 irregular petals forming a tube
Mint family

This is actually a white flower but because they are surrounded by prominent purple sepals it appears purple. So here we are with another color anomaly in the on-going saga of where do we feature it? For this guide—if it looks purple then purple it is. It is a common wildflower all through the mountains of northern California, southern Oregon, and on to the coast. You will find it in low to mid-elevation wet meadows where it grows to 6' tall. The white flowers are long and tubular, protruding beyond the deep lilac colored sepals, with prominent stamens and anthers. They bloom in dense 3" to 6" spikes that have a decidedly ragged look. The plant has a strong odor. I took these photographs (including the pale swallowtail on a Nettleleaf horse mint flower)in a meadow on Mt. Ashland in Southern Oregon.

ONE-SEEDED PUSSYPAW
Calyptridium monospermum
Miner's lettuce family

PUSSYPAW
Calyptridium umbellatum
Miner's lettuce family

NETTLELEAF HORSE MINT
GIANT HYSSOP
Agastache urticifolia
Mint family

RED BUSH BEARDTONGUE, REDWOOD PENSTEMON

Keckiella corymbosa 5 irregular petals forming a tube

Plantain family

The best place to see this gorgeous red *Keckiella* is along Canyon Creek Road on the steep rocky cutbanks close beside the roadway or on the dry, rocky flats just east of Junction City above the Trinity River. It loves hot, dry, fully exposed sites and is usually in full bloom when other flowers have long given up the ghost and gone into summer dormancy. The flower tubes are bright red and upwards of 2″ long. The lower lip is deeply incised and folded down at the tip while the upper petals form a hood extending out over the lower one. The entire plant, including the flowers, is hairy and sticky. I photographed these near the turn-off to Helena and the North Fork of the Trinity River along Highway 299 West in mid-July.

NOTHOCHELONE, TURTLEHEAD, WOODLAND PENSTEMON

Nothochelone nemorosa 5 irregular petals forming a tube

Plantain family

Nothochelone is found throughout the northern parts of the Klamath Mountains in mid- to high-elevations. I've found that it favors shaded sites typically found on the north sides of mountain peaks around rocky outcrops and scree slopes. Flowers vary from white, to deep yellow, brick red, pink, lilac and deep purple–truly a visual feast. The predominant colors, however, is white, brick red and deep purple which makes it exceedingly difficult to place it in any particular color regime. But, here it is safely ensconced in the red to purple section. All parts of this plant, including the flowers, are covered in dense sticky hairs. The deep green leaves are egg shaped near the base but lance shaped farther up the flowering stalk and all are toothed. The genus name comes from the Greek word for "false turtle" which refers to the corolla that looks like a turtle's head. I've photographed lots of turtles throughout the years but I fail to see the resemblance. I took this photograph on the ridge between Big Duck and Eaton Lakes in mid-July but you can find them blooming well into September in the scree slopes above Big Blue.

COPELAND'S OWL'S-CLOVER

Orthocarpus cuspidatus subsp. *copelandii* 5 irregular petals forming a tube

Broomrape family

This pretty little owl's-clover blooms all summer long in the high country, preferring dry, exposed, rocky meadows and openings. The plants are loosely branched but strongly upright with reddish stems that can reach to 12″ in height. The flowers are deep rose with a white and inflated lower lip. They tend to be gregarious and can cover large areas. I photographed these in Bear Basin in the Trinity Alps in late August. Other places to look for them are in Elderberry Duff and Parker Meadows in the Trinity Alps and Horse Heaven Meadows on the Trinity Divide.

SISKIYOU MOUNTAINS OWL'S-CLOVER

Orthocarpus cuspidatus subsp. *cuspidatus* 5 irregular petals forming a tube

Broomrape family

Probably the most distinguishing feature of this owl's clover is the white and inflated lower lip. The other petals are a pinkish-magenta. Unlike Copeland's, in which the flowers grow in a long tube shaped cluster, this one is shorter, flaring at the center, to give the flowering head a triangular shape. It has similar tastes in habitat, preferring relatively dry, open areas at mid- to higher-elevations. I shot these photographs near Stonewall Pass in the Trinity Alps and on Mt. Ashland in Southern Oregon. Look for it in places like High Prairie and Mary Blaine Meadow in the Trinity Alps and along the upper reaches of the Paynes Lake Trail in the Russians all summer long.

RED BUSH BEARDTONGUE, REDWOOD PENSTEMON
Keckiella corymbosa
Plantain family

NOTHOCHELONE
Nothochelone nemorosa
Plantain family

COPELAND'S OWL'S-CLOVER
Orthocarpus cuspidatus subsp. *copelandii*
Broomrape family

SISKIYOU MOUNTAINS OWL'S-CLOVER
Orthocarpus cuspidatus subsp. *cuspidatus*
Broomrape family

APPLEGATE'S INDIAN PAINTBRUSH
Castilleja applegatei — 5 irregular petals forming a tube
Broomrape family

This paintbrush is one of the most common species at middle to higher elevations throughout these mountains. You will find in bloom in dry, open, and exposed sites like rocky outcrops and ridgetops. It is both gregarious and found in small, scattered groups. Small clusters of flowers group at the tops of fairly long stems with leaves scattered along the stems. I photographed the red and yellow flowers above Upper Albert Lake in the Russians in late September. Yellow flowers are not common.

SCARLET INDIAN PAINTBRUSH
Castilleja miniata subsp. *miniata* — 5 irregular petals forming a tube
Broomrape family

This is the giant of all the paintbrushes in this area, reaching heights of three feet or more with large, bright red flowering heads that bloom throughout the summer into fall. It favors wet, open, and sunny slopes and is a trailside delight. Because of its size and its habitat, you will not confuse this paintbrush with any other in these mountains. I photographed this one in Long Canyon Meadows in late September.

FROSTED INDIAN PAINTBRUSH
Castilleja pruinosa — 5 irregular petals forming a tube
Broomrape family

Trying to identify the variety of *Castilleja* species in the region is a difficult task. There are too many species to to illustrate here—that's what the gigantic eight pound botanical manuals are for! This is one of the easy ones, however. It grows low to the ground and found in abundance blooming in places like Union Meadows in late May and June. I've found hundreds of them in that location and none of them over 7" tall. I photographed these along the Union Lake cut-off trail to Union Lake in the Trinity Alps in early June.

SPLITHAIR INDIAN PAINTBRUSH
Castilleja schizotricha — 5 irregular petals forming a tube
Broomrape family

This is the soft and fuzzy paintbrush—the entire plant is covered with wooly white hairs. Most are 3" to 6" in height with narrow three-lobed leaves scattered along the upright stems. The flowers are pale to dusty red and the stems are gray-green. It favors dry, open, and sunny sites. It is common to these mountains but good places to look for it are above Black's Basin in the Trinity Alps and below Twin Lakes in the Russians. I found a nice population just below Dogwood Lake in the Marble Mountain Wilderness which is where I shot this photograph.

APPLEGATE'S INDIAN PAINTBRUSH
Castilleja applegatei
Broomrape family

SCARLET INDIAN PAINTBRUSH
Castilleja miniata subsp. miniata
Broomrape family

FROSTED INDIAN PAINTBRUSH
Castilleja pruinosa
Broomrape family

SPLITHAIR INDIAN PAINTBRUSH
Castilleja schizotricha
Broomrape family

LITTLE ELEPHANT HEADS, BABY ELEPHANT HEADS
Pedicularis attollens
Broomrape family

5 irregular petals forming a tube

This wildflower definitely classifies as "cute." The tiny light to dark pink flowers really do look like little elephant heads. The upper petals are narrow and recurved upward from a protruding bulge that resembles an elephant's trunk—while the lower petals spread away and droop forming the ears. There are many flowers in a compact head blooming near the top of a leafless stalk. Color ranges from white for the newly opened flowers to deep pink when older . Leaves are mostly basal, linear, and finely divided. Their preferred habitat is wet meadows, fens, and stream banks at mid- to high-elevations. I shot this photograph in Parker Meadows in mid-July.

INDIAN WARRIOR, WARRIOR'S PLUME
Pedicularis densiflora
Broomrape family

5 irregular petals forming a tube

Indian warriors bloom early at low elevations, often competing with milkmaids and manzanita to be the first to show. I've found them in full bloom in mid-January in some years but, more typically in late February. Low elevation warriors are larger and stand much taller than those you will find in the high country. Lower down they reach heights of 12" or more but higher up rarely exceed 6". The blooms and foliage are pretty much the same however the flowers are a deep vibrant red or orange and the leaves long and deeply dissected—almost fern-like. I shot this photograph along Scorpion Creek below Bonanza King Lookout in early June, but look for them along the Swift Creek trail as well. **Note**: I found and photographed a group of beautiful, buttery-yellow warriors (see inset) along Highway 36 west of the town of Platina. Life is full of surprises!

LEAFY LOUSEWORT, PARROT'S BEAK, SICKLETOP LOUSEWORT
Pedicularis racemosa
Broomrape family

5 irregular petals forming a tube

This pretty little wildflower is found throughout the Trinities, the Russians and the Marbles at mid- to high-elevations especially around water. The leaves are undivided, lance-like, and finely toothed. While the flowers in this photograph appear mostly pink, the bulk of those surrounding this group were white. I shot this photograph below the outlet to Lower Wright Lake in the Marble Mountain Wilderness in late July.

MEXICAN HEDGENETTLE
Stachys mexicana
Mint family

5 irregular petals forming a tube

This wildflower is an eye catcher -—beautiful deep magenta with white stripes and spots emanating from the throat on the lower lobe. At first glance, they resemble penstemon flowers. Overall, though, the square stems are strongly upright and the sharply ovate leaves are covered with soft hairs. This species prefers damp shaded sites. It is widespread throughout these mountains and especially abundant on the coast. I photographed these at the head of Morris Meadows in late July and have also seen them on Scott Mountain and along Forest Route 17 below the Parks Creek Divide.

LITTLE ELEPHANT HEADS
Pedicularis attollens
Broomrape family

INDIAN WARRIOR, WARRIOR'S PLUME
Pedicularis densiflora
Broomrape family

LEAFY LOUSEWORT, PARROT'S BEAK
Pedicularis racemosa
Broomrape family

MEXICAN HEDGENETTLE
Stachys mexicana
Mint family

FOXGLOVE
Digitalis purpurea — 5 irregular petals forming a tube
Plantain family

In this area we usually associate foxgloves with coastal environs, but there is a small population found along the North Fork of Coffee Creek. It is a non-native that is highly poisonous to livestock as well as humans. I have a hunch that the folks who built their homes (think Hodges Cabin) along the creek in the early 20th century probably planted foxgloves in their gardens and that those now growing in the area are garden escapes. That being said, it is a beautiful flower that should be mentioned here. It grows up to 6′ or 7′ tall with deep rose-purple (sometimes white), tubular shaped flowers clustered along the upper portions of the stem. The throat of each flower is speckled with white and deep purple. They usually bloom in July.

BREWER'S MONKEYFLOWER
Erythranthe breweri — 5 irregular petals forming a tube
Lopseed family

I believe this is one of the tiniest monkeyflowers you will find in these mountains. The plants hug the ground, hardly ever getting over 2″ tall. The pink to deep rose, tube shaped flowers are minute, most barely 1/8″ across, but they bloom in dense colonies in open and sandy meadow areas at mid- to high-elevations. A close-up reveals classic monkeyflower structure with recurved upper petals and three lobed petals that form a lower lip. The lower petals are yellow at the throat while the upper petals are deep magenta at the throat. All the petals are pale pink to near white at the notched tips. Look for them in places like Doe Flat (where I photographed these), along the PCT between Doe and Little Granite Lakes, or the dry meadows of Horse Creek in the Trinity Alps from mid-summer to mid-fall.

CRIMSON MONKEYFLOWER, SCARLET MONKEYFLOWER, CARDINAL MONKEYFLOWER
Erythranthe cardinalis — 5 irregular petals forming a tube
Lopseed family

This is the most brilliantly colored of all the monkeyflowers in this area. The large flowers are a beautiful cardinal red with recurved petals streaked with buttery-yellow at the throat. They bloom at the ends of tall, upright stalks that can reach 24″ in height. Stalks, leaves, and sepals are all covered with sticky hairs. Look for them in moist, sandy areas along the Trinity River and around springs and seeps at lower elevations. Along the river they are often partially submerged by rising spring run-off. I have found large populations growing in association with horsetails. They can bloom all summer long if conditions are right. I photographed these near the mouth of Indian Creek on the Trinity River in late July.

LEWIS' MONKEYFLOWER
Erythranthe lewisii — 5 irregular petals forming a tube
Lopseed family

This is a large, beautiful, and showy monkeyflower. It graces areas where there is running water, the shorelines of lakes, and the edges of seeps and snowmelt ponds early in the blooming season which, in the Trinity Alps, can mean anywhere from early July to late August. The brilliant pink flowers are 3/4″ in diameter on plants that are 12″ to 24″ in height. Because of their size and color, they cannot be confused with any other monkeyflower in this area. I photographed these in the steep, wet meadows at the back of Grizzly Lake in August but other good places to look for them is in the upper bowl of Long Canyon and around the shoreline of Lake Luella.

FOXGLOVE
Digitalis purpurea
Plantain family

BREWER'S MONKEYFLOWER
Erythranthe breweri
Lopseed family

CRIMSON MONKEYFLOWER
Erythranthe cardinalis
Lopseed family

LEWIS' MONKEYFLOWER
Erythranthe lewisii
Lopseed family

PURPLE MOUSE-EARS, DOUGLAS' MONKEYFLOWER, BROWNIES

Diplacus douglasii 5 irregular petals forming a tube

Lopseed family

This tiny monkeyflower often hides in plain sight. The 3" to 4" plants like dry, exposed rocky sites at lower elevations where they bloom in profusion. A solitary magenta-purple flower opens atop a short stalk that rises from a basal rosette of light green and deeply veined leaves. The upper two petals of the flower are the largest and give rise to one of its best common names—purple mouse-ears. The lower petals are much shorter. They should not be confused with dwarf monkeyflower described below and can be distinguished by the broad inflated throat at the base of the petal lobes in this species. I photographed these alongside the road to the Big French Creek Trailhead in the Trinity Alps in late June.

KELLOGG'S MONKEYFLOWER

Diplacus kelloggii 5 irregular petals forming a tube

Lopseed family

This tiny magenta-purple monkeyflower resembles the Douglas' monkeyflower except that the flowers lack the large upper petals and have well developed lower petal lobes. Two or three flower bearing stems rise out of the basal leaf cluster, each stem with a solitary flower at the top. The flower is tubular with the petals opening away from the throat. The throat itself is yellow and spotted with magenta. The leaves are more olive-green with reddish tinted veins and purple underneath. I photographed this one along the Parks Creek Road just off Highway 3 in mid-June.

DWARF MONKEYFLOWER

Diplacus nanus var. *nanus* 5 irregular petals forming a tub

Lopseed family

This really is a dwarf wildflower, barely over two inches tall and easy to miss while trudging uphill on a hot day under the weight of a loaded pack. My observant granddaughter Cristan first found them on a trip a few years ago and brought them to my attention. The beautiful deep pink corolla is almost trumpet shaped with petals that flare away from the center exposing a deep yellow throat spotted with pink. The compact leaves are olive-green, covered with minute hairs and narrowly oval. I shot this photograph near the summit of Scott Mountain in late June and the inset above West Boulder Lake in mid-July.

LAYNE'S MONKEYFLOWER

Diplacus layneae 5 irregular petals forming a tube

Lopseed family

This species is so small and delicate that a backpacker or hiker might never see it. The long, tube shaped flowers are a beautiful pink with a white and yellow throat spotted with a deeper magenta. Petals fan away from the throat in an almost perfect circular shape. Leaves, stems, and sepals are all hairy. I've never seen it much over 3" tall and always in huge colonies growing in dry, exposed, sandy sites. One of my grand-daughters discovered it quite by accident on a pack trip to Log Lake in August—which is where I photographed these. They were growing in light colored granitic sand and very hard to see.

PURPLE MOUSE-EARS
Diplacus douglasii
Lopseed family

KELLOGG'S MONKEYFLOWER
Diplacus kelloggii
Lopseed family

DWARF MONKEYFLOWER
Diplacus nanus
Lopseed family

LAYNE'S MONKEYFLOWER
Diplacus layneae
Lopseed family

DAVIDSON'S PENSTEMON
Penstemon davidsonii
NEWBERRY'S-DAVIDSON'S PENSTEMON HYBRID
Penstemon newberryi X davidsonii
NEWBERRY'S PENSTEMON, MOUNTAIN PRIDE
Penstemon newberryi 5 irregular petals forming a tube
Plantain family

For ease of comparison I've chosen to include all of the Newberry's and Davidson's variations on this page, including the blue and bluish-purple varieties.

Newberry's is a highly variable penstemon with tubular shaped flowers ranging in color from electric pink to deep magenta with yellow spots around the flower opening. Long hairs decorate the throat. Its leaf structure varies as well because it hybridizes with Davidson's penstemon. Newberry's tends to be a bit rangy with larger, leathery leaves that are toothed at the margins while Davidsons often forms dense mats with smaller leaves that are likewise toothed. Davidson's flowers are tubular, small, and soft mauve on the outside and lighter in the throat which is trimmed with white hairs. Davidson's and the Newberry's/Davidson's hybrid both lack the yellow spots found on Newberry's. Newberry's is found throughout a variety of habitats but almost always around rocks and rocky outcrops at mid- to high-elevations. The Davidson's I've found have, for the most part, been occupants of higher elevation exposed ridgetops. Flowers of the hybrid version are light to dark blue sometimes tinged with a bit of magenta and, in shape, more closely resembling Newberry's. The hybrid are found together with Newberry's and I have found a mix of pink and blue flowers on the same stalk. I photographed the vibrant pink Newberry's above the shoreline of Doe Lake in the Trinity Alps. The photograph of the Davidson's was taken on the ridgetop between Trail Gulch and South Fork Lakes in the Trinity Alps. I photographed the blue hybrid on the ridge above Upper Albert Lake in the Russian Wilderness. They bloom from July into September.

PRIDE OF THE SIERRAS
Penstemon newberryi var. *newberryi* 5 irregular petals forming a tube
Plantain family

Recognized more by its color than anything else (by amateurs anyway) this penstemon is a real standout. Except for color it closely resembles its cousin Newberry's penstemon. Newberry's, however, is typically a lighter shade of pink and not the deep, electric rose-red of this one. The plants of both are similar—forming low growing evergreen mats with deep, leathery green leaves that are toothed at the margins. Both inhabit dry, rocky places such as granite outcrops and crevices where, as one Yosemite naturalist notes, "it looks impossible for flowers to grow." I found and photographed these blooming in serpentine outcrops along the Pacific Crest Trail below Deadfall Lakes on the Trinity Divide in early August.

DAVIDSON'S PENSTEMON
Penstemon davidsonii var. *davidsonii*
Plantain family

DAVIDSON'S-NEWBERRY'S PENSTEMON HYBRI
P. newberryi X *davidsonii*
Plantain family

NEWBERRY'S PENSTEMON
Penstemon newberryi
Plantain family

PRIDE OF THE SIERRAS
Penstemon newberryi var. *newberryi*
Plantain family

PINK-FLOWERED STICKSEED, PINK STICKSEED
Hackelia mundula
Borage family

5 symmetrical petals, ovary superior

This stickseed has a limited distribution in this area. Larger populations occur in the Sierra Nevada—especially in the southern Sierra. In July, i've found it in small, isolated colonies around the edges of wet meadows, near springs, and the banks of streams along the Pacific Crest Trail above West Boulder Lake and along the trail between South Fork Lakes and Saloon Creek (which is where I photographed these). The beautiful soft pink flowers are large, up to ½" across, and bloom in loose clusters at the tops of tall stems. Plants can reach heights of 3'. It is the only pink stickseed in these mountains and finding is a real plus in anybody's wildflower book!

WHORLED MILKWEED, NARROW-LEAF MILKWEED
Asclepias fascicularis
Dogbane family

5 symmetrical petals forming a tube

I really like milkweeds because they are beautiful and showy. You can find them almost everywhere at lower elevations in these mountains. This species is tall and, at first glance, gangly—because of the narrow leaves. The thick bunches of small purple and white flowers are every bit as impressive as those of the other two species illustrated in this guide (showy milkweed below and purple milkweed pg. 279). Monarch butterflies are fond of milkweeds as well—in fact it is their main food source. For that reason, it is important not to disturb populations of any of these species. Milkweeds like disturbed places alongside roads and flats. They are easy to spot even as you are driving. I photographed these alongside Trinity Mountain Road above Lewiston Lake and along Highway 3 near Douglas City in mid-June.

SHOWY MILKWEED
Asclepias speciosa
Dogbane family

5 symmetrical petals forming a tube

This milkweed is common along the edges of roads and on dry, open flats throughout this area. This is the largest of the milkweeds, with some plants approaching 4' in height—sporting several globose heads of large beautiful pink and deep magenta flowers. Flower petals are sharply linear. The leaves are large, fuzzy, and gray-green. A sticky, milky sap appears when stems are broken. It is hard to miss this one. I shot this photograph just above Lewiston Lake in late June.

PINK-FLOWERED STICKSEED
Hackelia mundula
Borage family

WHORLED MILKWEED
Asclepias fascicularis
Dogbane family

SHOWY MILKWEED
Asclepias speciosa
Dogbane family

BITTER DOGBANE
Apocynum androsaemifolium 5 symmetrical petals forming a tube
Phlox family

Dogbane is a ground hugging plant that inhabits open, sunny, rocky slopes and flats. Leaves are long, opposite, and oval---dark green above and pale beneath. In late summer, at higher elevations, the entire plant turns a brilliant red while lower down it tends to turn more yellow. The pink to white tubular shaped flowers resemble those of manzanita and bloom in thick bunches. Like milkweed, dogbane exudes a sticky, milky sap when its stems are broken. I shot these photogtaphs near Bee Tree Gap near the Long Canyon trail in the Trinity Alps in July. It is common throughout these mountains.

VARIABLE-LEAF COLLOMIA
Collomia heterophylla 5 symmetrical petals forming a tube
Phlox family

This beautiful little collomia is anything but shy. The small, tubular, brilliant pink flower petals flare away from the centers of white and yellow, and bloom en-masse from tufts of leaves at the tops of short, branched stems. Stamens and anthers are safely hidden deep in the tube. The entire plant is hairy to some degree but the sepals surrounding the flower tubes are exceedingly so. Though only a few inches tall, the color is so intense that you cannot miss them as you walk or even drive along. The lobed leaves distinguish this from other annual collomias. Large colonies of them bloom from May into June in sandy soil alongside low elevation trails throughout these mountains. I photographed these along the Boulder Creek Trail in the Trinity Alps in late May.

TRACY'S COLLOMIA
Collomia tracyi 5 symmetrical petals forming a tube
Phlox family

I found this little collomia blooming all along the trail to the top of Packer's Peak in the Trinity Alps in late June. There were hundreds of tiny bunches growing in the direct sun, on the open, exposed slopes in serpentine soils. Flowers range from white to lavender. The petals were streaked with darker lavender with some slight white edging. They were all about ¼" in diameter or smaller. Pollen on the anthers is bright blue. It is a multi-branched plant with several reddish-purple stems crowded wth lance shaped, hairy leaves. It was hard to avoid stepping on some of these as they were growing right in the middle of the trail–which is not surprising as the trail is becoming not-existent these days from lack of use.

BITTER DOGBANE
Apocynum androsaemifolium
Dogbane family

VARIABLE-LEAF COLLOMIA
Collomia heterophylla
Phlox family

TRACY'S COLLOMIA
Collomia tracyi
Phlox family

NORTHERN GENTIAN, AUTUMN DWARF GENTIAN
Gentianella amarella subsp. *acuta* 5 symmetrical petals forming a tube
Gentian family

This beautiful little member of the gentian family is widespread throughout these mountains and blooms summer to fall. It is found in marshy meadows and other wet areas where it grows to 12" to 15" in height. The delicate mauve to light pink flowers bloom on short stalks rising out of the leaf axils. Five petals form a tube, are rounded at the tips, and flare away above the sepals exposing a prominent white fringe surrounding the ovary, pistils, and stamens. The basal leaves are spoon-shaped. Numerous lance shaped leaves crowd upward along the stems in opposite pairs. The flowers can also be white but I've yet to find any. I photographed these in Parker Meadow in the Trinity Alps in mid-September. I've also found them in the meadows below East Boulder Lake in the Trinity Alps, in the meadows below Lake Eleanor (just outside the Trinity Alps), and in the meadows along Parks Creek Road on the Trinity Divide.

SCARLET GILIA, SKY ROCKET, SCARLET TRUMPET
Ipomopsis aggregata 5 symmetrical petals forming a tube
Phlox family

Sky rocket is the perfect name for this brilliant red wildflower. The long, funnel shaped flowers flare sharply away at the tips revealing long stamens topped with bright yellow anthers—which make them look like exploding fireworks. The petals are lightly spotted with yellow near the throat. It can be found from mid- to high-elevations wher it prefers exposed to semi-shaded, dry, sandy sites. It grows in great profusion along the upper portions of the trail between Syphon and Russian Lakes in the Russian Wilderness where I shot the small photograph, and along the upper reaches of the Stonewall Pass Trail near Red Mountain Meadow where I shot the large photograph. **Note:** In some areas it is called scarlet honeysuckle because of the drops of sweet nectar that can be sucked out of the basal end of the flower.

VARIABLE LINANTHUS
Leptosiphon parviflorus 5 symmetrical petals forming a tube
Phlox family

The most beautiful display of this little wildflower I've ever seen was on steep banks along the road leading up to Hirz Mountain Lookout in late April. In one area it covered an entire hillside—blooming with Kellogg's monkeyflowers and pink grass. It was quite striking. Since that time, I've found it in several other areas including alongside trails above Shasta lake and along Highway 299 above the Trinity River Gorge. Flowering stems can be several inches tall with whorls of needle like leaves scattered along. A thick mass of leaves encircles the stems just below the two or three thin stems that support the flowers. Flowers range from white to deep pink with a white band at the center and are solitary atop the stems. Deep yellow or orange stamens and anthers sit deep within the center of the tube. I shot these photographs on Hirz Mountain in April.

NORTHERN GENTIAN
Gentianella amarella subsp. *acuta*
Gentian family

SCARLET GILIA, SKY ROCKET
Ipomopsis aggregata
Phlox family

VARIABLE LINANTHUS
Leptosiphon parviflorus
Phlox family

MIDGET PHLOX, SLENDER PHLOX
Microsteris gracilis
Phlox family

5 symmetrical petals forming a tube

My first encounter with this tiny phlox was on the steep cutbanks and cliffs alongside Canyon Creek Road in mid-April. They were scattered throughout the area and numerous in shade and sun. The petals are delicate, lavender-pink, with darker areas of purple near the throat that surrounds the greenish-white center of the corolla. The plants are small (2″ to 4″ tall) and the flowers about 1/8″ across. The plants I photographed were multi-branched with several flowers on each plant. You really have to look for this one as it tends to hide down in and amongst other plants, especially heron's bill which, from a distance, the flowers resemble. The entire plant, with the exception of the flowers, is hairy.

SPREADING PHLOX
Phlox diffusa
Phlox family

5 symmetrical petals forming a tube

This was my mother's favorite high country flower and through the years it has remained one of mine as well. It's common and blooms in large colonies all through these mountains in open, sunny areas popping up as the winter snows melt. The flowers are highly variable with regards to color—ranging from white to pink to purple. Flowers crowd the plants and fade to white as they age. The mats of leaves are needle-like with the flowers blooming in low mounds over the tops. Good places to look for them in late spring and early summer are in the area between Union Meadows, and the Parker Divide (all along the Swift Creek Trail) and in the basin surrounding Gray Rocks Lakes on the Trinity Divide. I shot the large photograph above Horse Creek in early July and the smaller one on Scott Mountain in mid-June. It is not unusual to still find blooms in late September in protected places at higher elevations.

SHOWY PHLOX
Phlox speciosa
Phlox family

5 symmetrical petals forming a tube

This is one of those wildflowers you will find blooming early in May in foothills around the Central Valley and through early June at mid-elevations. The large, pale to deep pink flowers grow in clusters on loosely branched plants that grow to 7″ tall and, in some instances taller. Sometimes individuals climb vine-like up into the low branches of surrounding brush. Look for them along the cutbanks of roads and highways, which always seems to be the best place to locate them. I photographed these along Highway 299 on the west side of Buckhorn Summit in June. Large populations of them thrive on Scott Mountain and are common along the road to the Tangle Blue Trailhead. It can be, quite literally, a show-stopper on any drive into the mountains—just find a safe place to pull over, especially on a major highway.

MIDGET PHLOX
Microsteris gracilis
Phlox family

SPREADING PHLOX
Phlox diffusa
Phlox family

SHOWY PHLOX
Phlox speciosa
Phlox family

HENDERSON'S SHOOTING STAR, MOSQUITO BILLS

Primula hendersonii

Primrose family

5 symmetrical petals forming a tube

This, the most common shooting star in our area, blooms mainly at lower elevations in the foothills of the surrounding mountains. They are also among the earliest bloomers, often showing up in late January or early February. The plants are usually 4″ to 7″ tall with loose clusters of deep magenta to light purple inverted flowers at the tops of reddish stems. As with all shooting stars, the flowers are easily identified and this one is especially easy because of its size and the rounded, oblong leaves. I've found it blooming along the lower stretches of the Swift Creek and Stewart's Fork Trails in May and early June. These photographs were taken on the road to the Tangle Blue Trailhead and on Bolt's Hill above Lewiston Lake.

SIERRA SHOOTING STAR, JEFFREY'S SHOOTING STAR
TALL MOUNTAIN SHOOTING STAR

Primula jeffreyi

Primrose family

5 symmetrical petals forming a tube

This is the giant among local shooting stars with stalks as tall as 24.″ You will find it in wet meadows and alongside streams, springs, and seeps. It likes its feet in the water! The only other shooting star that approaches this in size is the Alpine shooting star shown here. The pale lilac flowers bloom in a loose cluster at the tops of the stems. Look for it in the wet meadows alongside the road to the Tangle Blue Trailhead, along the Stoddard Lake Trail in the Trinity Alps, and in the meadows around Big Carmen Lake on the north side of Scott Mountain—which is where I photographed these.

ALPINE SHOOTING STAR

Primula tetrandra

Primrose family

5 symmetrical petals forming a tube

The size of this shooting star and its vibrant pink blooms make identification easy—that and the fact that it is only found in fairly wet meadows and open areas at higher elevations in these mountains. I've found it as tall as 20″ but most often seems to be in the 7″ to 12″ range. I photographed this one near the outlet of Lois Lake in the Trinity Alps, but have found it around Little Bear Lake, in the meadows around Upper Caribou Lake, and in rocky areas around the Duck Lakes in the Russians.

HENDERSON'S SHOOTING STAR
Primula hendersonii
Primrose family

SIERRA SHOOTING STAR
Primula jeffreyi
Primrose family

ALPINE SHOOTING STAR
Primula tetrandra
Primrose family

Sierra shooting stars decorate a Darlingtonia fen along with Water plantain buttercup on Scott Mountain in early June.

SIERRA PRIMROSE
Primula suffrutescens 5 symmetrical petals forming a tube
Primrose family

This is one of our true alpine wildflowers where it is found growing among the rocks on scree slopes and along the bergschrund line on rocky shelves beneath the tallest peaks around Papoose, Caribou, and Grizzly Lakes. In my years in the high country I have never encountered it anywhere else, although Marla Knight has mentioned sighting it in rocks alongside the New Caribou Lakes Trail below Little Caribou Lake. The flowers are unmistakable—five deep magenta pink petals surrounding a yellow center blooming on stems up to 7" tall. The stiff leaves form a rosette, appear wrinkly, are elongated, and toothed on the margins. I photographed these just below the granite peaks above Caribou Lake Lake in mid-August.

HAIRY-FRUIT VALERIAN, MOUNTAIN VALERIAN
Valeriana pubicarpa 5 symmetrical petals forming a tube
Valerian family

I found this gorgeous little valerian blooming in decomposed granite at the top of the ridge between Little Duck Lake and the Sugar Creek drainage. All of the populations were in shaded sites on the west and northwest facing parts of the ridge. Typically, the plants were no more than 12" tall with deep green, round, basal leaves with smaller, opposite, lance shaped leaves scattered farther up the flowering stalk. Beautiful light pink to magenta flowers bloom in clusters along the upper 1/3 of the stalk with prominent stamens. I photographed this one at the top of the ridge above Little Duck Lake in August.

TWIN FLOWER
Linnaea borealis var. *longiflora* 5 petals joined into a bell
Twin flower family

Twin flowers inhabit shaded, damp forest floors—places where you think you would be running into faeries or gnomes. The plants are small, rarely exceeding 6" in height, and form dense, ground covering mats when conditions are favorable. The top of the central stem gives rise to two smaller stems that emerge upward at an almost perfect 45° angle with two delicate, pink and white flowers hanging at another 45° angle outward from the end of each. The pendant flowers are bell-shaped and about 3/8" long. A prominent white pistil extends beyond the petals and forms a perfect little clapper to the bell of the flower. I shot the large photograph along Mule Creek at the foot of the Trinity Alps and the smaller photograph along the Boulder Creek Trail—both in mid-July. There is a section of the Stewart's Fork Trail, near the upper end of Oak Flat where these little beauties can be found en-masse. **Note:** The genus is named after the 'father' of modern plant taxonomy, Carl von Linne, also known as Linnaeus. The same species grows in Sweden, where he lived and worked.

SIERRA PRIMROSE
Primula suffrutescens
Primrose family

HAIRY-FRUIT VALERIAN
MOUNTAIN VALERIAN
Valeriana pubicarpa
Valerian family

TWIN FLOWER
Linnaea borealis var. *longiflora*
Twin flower family

ROSY PUSSYTOES, ROSY EVERLASTING
Antennaria rosea — Flowers in composite heads
Sunflower family

The deep green leaves of this pretty little pussytoe are longer, more strongly upright, stiff, and not nearly as fuzzy as other *Antennaria* species. They also clasp the flowering stalk along its entire length right to the base of the flowering head. Like all pussytoes, the little flowers bloom in a compact head at the tops of the stems, but these are graced with a beautiful rose-pink mass of ray flowers surrounding a Q-tip-like white button of disc flowers top and center. I've never encountered many of these on my high country adventures but I did find and photograph these on a rocky ledge just above Doe Lake in the Trinity Alps in mid-September and along the road to Horse Flat Campground above the Trinity River in early May.

BREWER'S SWAMP THISTLE
Cirsium douglasii var. *breweri* — Flowers in composite heads
Sunflower family

This is a common native thistle found in damp environments at mid- to high-elevations throughout this area. It is a slender-stalked thistle covered with white wooly hairs that grows to about 3′ in height. The lavender to pink flower heads are about 1″ long and bloom in small clusters. It is spiny and, unless you enjoy excruciating pain, you should never reach out and grab one. Spines occur along the stems, leaf margins and the bracts below the flowers. I photographed this one in Elderberry Duff in the Trinity Alps in late August.

SNOWY THISTLE, RED THISTLE
Cirsium occidentale var. *candidissimum* — Flowers in composite heads
Sunflower family

This is probably the showiest of all the thistles found in these mountains and you don't even need to get out of your car to find them. They are a roadside attraction from low- to mid-elevations where they are frequently seen on exposed, dry, rocky slopes. The entire plant including stems and leaves are cobwebby white, almost silvery. The large flower heads are an intense and striking red. They cannot be confused with any other thistle in this area. I photographed this one along Highway 3 near Callahan in August.

BULL THISTLE
Cirsium vulgare — flowers in composite heads
Sunflower family

This is a common, non-native thistle found from the Klamath Mountains all the way to the coast. It is tall and gangly with stems spreading haphazardly from a central stalk. Every part of the plant is covered with spiny prickles which can be quite painful if you inadvertently brush up against them or unknowingly grab a handful. Stems are reddish-green. The olive-green leaves are found all along the stems and are long and sharply lobed almost all the way to the mid-vein. Flower heads are numerous and range from pink to deep bluish-purple. Bull thistle was introduced from Europe and is now found across North America alongside roads and in other disturbed areas. I photographed these along Highway 3 on Scott Mountain in mid-August.

ROSY PUSSYTOES, ROSY EVERLASTING
Antennaria rosea
Sunflower family

BREWER'S SWAMP THISTLE
Cirsium douglasii var. *breweri*
Sunflower family

SNOWY THISTLE, RED THISTLE
Cirsium occidentale var. *candidissimum*
Sunflower family

BULL THISTLE
Cirsium vulgare
Sunflower family

PHILADELPHIA FLEABANE

Erigeron philadelphicus

Flowers in composite heads

Sunflower family

This is a common fleabane that occurs on open, rocky sites, along streams, and in other damp places throughout the mid-elevations of the area covered by this book. It can grow in huge colonies. The flower heads can be up to 1" across with a multitude of long, pinkish to blue slender ray flowers surrounding each compact head of yellow disc flowers. A single stem arising from the rootstock loosely branches midway to support several flower heads. Leaves to 6" long clasp the stalks. I photographed these in Long Canyon Meadows in the Trinity Alps.

WESTERN PEONY, BROWN'S PEONY

Paeonia brownii

Flowers in composite heads

Peony family

This is the sole member of the Peony family found in this area. While I've not encountered it in Trinity County there are isolated populations in the Weaverville, Douglas City and Hayfork areas. The plants are bushy with several stems rising from a common rootstock averaging between 12" and 20" in height. The leaves are divided with oblong, rounded leaflets that—because of a waxy powder—have an bluish appearance, especially in the older leaves. One or more flowers nod at the ends of tall, thick red stems with spreading, deep reddish-green to purplish sepals edged with yellow-gold. Sepals and petals appear much the same. Fleshy, elongated pistils are surrounded by numerous stamens. I photographed this one along Highway 97 east of Weed in early May

WESTERN STAR FLOWER

Lysimachia latifolia

many petals

Myrsine family

These little beauties love the deep shade and lingering spring moisture provided by an overhanging forest canopy. They are common throughout the entire area often blooming in large populations with flowers that really do look like little stars. Plants are about 6" tall, with five leaves in a whorl around a strong upright stem. Each stem has between one and three bright pink flowers with overlapping, sharply pointed petals and white stamens topped with deep yellow anthers. The flowers fade to white as they age. I photographed these along the Swift Creek Trail but you will find them along any of the lower elevation trails like Stewart's Fork, or in the foothills.

PHILADELPHIA FLEABANE
Erigeron philadelphicus
Sunflower family

WESTERN PEONY
BROWN'S PEONY
Paeonia brownii
Peony family

WESTERN STAR FLOWER
Lysimachia latifolia
Myrsine family

Note about lewisias: there are four varieties of *L. cotyledon* found in these mountains; var. *heckneri* (pictured on the opposite page) has sharply toothed leaves, var. *cotyledon* has leaf margins that are entire, and var. *howellii* has wavy-edged leaves. The flowers of all three are so similar as to be almost indistinguishable. A fourth hybrid variety is *Lewisia X whiteae*.

CLIFF MAIDS, SISKIYOU LEWISIA
Lewisia cotyledon many petals
Miner's lettuce family

As noted above, the leaves of this variety are entire—not toothed and not wavy. They form a loose basal rosette, are long, and tend to be more olive-green. The white and pink striped flowers bloom in loose bunches along the more or less upright stems and are virtually indistinguishable from those of v. *heckneri*. I shot several photographs of this variety on the talus slopes above Upper Albert Lake and Big Blue both in early September in the Russians.

HECKNER'S LEWISIA
Lewisia cotyledon var. *heckneri* many petals
Miner's lettuce family

As noted above, the leaves of this beautiful lewisia are sharply toothed, large, oval to almost round , bright green, and crowded in a basal rosette. This variety is especially striking as it graces mostly shaded mid-stream rocks and other rocky outcrops directly above the surging springtime waters of Stewart's Fork. The candy-striped pink and white flowers are large and numerous, blooming in loose bunches at the tops of long, reddish stems. I photographed these in mid-June.

QUILL-LEAVED LEWISIA, LEE'S LEWISIA
Lewisia leeana many petals
Miner's lettuce family

The deep pinkish-purple flowers of Lee's lewisia are tiny but bloom in large clusters at the tops of branched stems, which make them easy to spot. The leaves form small basal tufts, are linear, fleshy, and pointed. It prefers sunny, open rocky areas. A great place to see this delicate lewisia is at Little Duck Lake in the Russian Wilderness. They bloom in abundance around the lake but they are common all through the high country. I photographed these in rocks above Lion Lake in the Trinity Alps in late May. Also pictured (in the inset) is a white variation of this lewisia which I found in large populations on the slopes above Salmon Lake in the Trinity Alps in early September.

LEWISIA COTYLEDON / LEWISIA LEEANA HYBRID
Lewisia X whiteae many petals
Miner's lettuce family

The narrow leaves of this hybrid have entire margins, are more upright in nature, and are somewhat succulent in appearance. The flowers are vibrant pink, 1/3" to 1/2" wide, with deep red stripes and are a bit smaller than those of the other cotyledon varieties—but much larger than Lee's. They bloom in loose clusters along the flowering stalks. Several stalks arise from each leaf cluster. They prefer rocky outcrops and talus slopes. I photographed these betseen Lower and Upper Albert Lakes in the Russians in September.

CLIFF MAIDS, SISKIYOU LEWISIA
Lewisia cotyledon
Miner's lettuce family

HECKNER'S LEWISIA
Lewisia cotyledon var. *heckneri*
Miner's lettuce family

QUILL-LEAVED LEWISIA, LEE'S LEWISIA
Lewisia leeana
Miner's lettuce family

LEWISIA HYBRID
Lewisia X whiteae (*L. cotyledon*/*L. leeana* hybrid)
Miner's lettuce family

WOODLAND WIRE LETTUCE
Stephanomeria lactucina
Sunflower family

Many petals

I look at these two photographs and see two different wildflowers but they are, indeed, the same species. I found them in late September within 60′ of each other in full bloom along the Deacon Lee Trail in the Russians. The basic plant structure is identical—with thin flowering stems about 24″ long trailing along the ground with the combined weight of leaves and flowers. Leaves of both are gray-green, long, and lance shaped. The flower heads on the left are a light violet or mauve fading to a creamy-white center with prominent cream colored anther tubes topped with dark mauve stigmas. Individual ray flowers are almost rectangular, squared and notched at the tips and rimmed with a thin creamy white band. The ones on the right are a dark reddish-purple fading to lighter purple at the center with a prominent anther tube topped with protruding purple stigmas. Individual ray flowers are a bit more elliptical, folded at the tips and sharply pointed. I found numerous, identical fuzzy white seed heads on both. **Note:** In 2014 the July Fire ravaged a large area of the Russians, including the stretch of the Deacon Lee Trail where I had previously found these. On a trip into Russian Lake in 2015 I found no sign of the plants or flowers—the entire area severely burned and thick with bracken fern. One has to hope that they make a come-back.

The Deacon Lee Trail is one to savor. It starts out high and, over its entire length, offers up gorgeous views of the Trinity Alps to the south, the Siskiyou, Marble and Coast Range to the west and north, and the Salmon River far below. It skirts the southern fringes of the Russian Wilderness and wanders through beautiful mixed conifer forests (a lot of which burned in the 2014 July Fire) and several wet and dry meadows. It's a wildflower hunters paradise from beginning to end but perhaps the most stunning time is in the fall when the expansive populations of sulphur flowered buckwheat turn orange and red. Below the trailhead and along the open, rocky ridge top, this display is a show stopper. The photograph on the left was taken on a warm morning in July and the one on the right during the onset of an intermittent snow storm in late September

WOODLAND WIRE LETTUCE
Stephanomeria lactucina
Sunflower family

The Deacon Lee Trail in the Russian Wilderness offers up plenty of wildflowers along it's entire length, including the Woodland wire lettuce pictured above, as well as great views of the surrounding mountains.

MOUNTAIN PINK CURRANT, SIERRA CURRANT

Ribes nevadense — 5 symmetrical petals - ovary inferior

Gooseberry family

This shrub is hard to miss when it's blooming. Reaching to 5', it has wrinkled leaves that resemble maple leaves. The flowers bloom in a brightly colored raceme of deep pink with slightly reflexed sepals surrounding white, tube shaped petals that hide the stamens and anthers. They are found throughout the mountains from low to mid-elevations in moist, shaded sites. The fruit of the currant is not stickery so cannot be confused with gooseberries, to which they are related. The deep, blue-black berries ripen in bunches and are almost always coated with a light blue to gray bloom. A similar species, *Ribes sanguineum*, is also present in the book's area. It differs in having larger flowers with more reflexed sepals. I photographed the flowers along the trail to Duck Lakes and the berries along the Deacon Lee Trail in September both in the Russians.

SIERRA GOOSEBERRY

Ribes roezlii — 5 symmetrical petals - ovary superior

Gooseberry family

You cannot mistake this one! The thorny shrub is loaded with pendulous, velvety red and pink flowers in the spring and sharp, spiny fruits in the summer and fall. It grows in dry sites along the edges of meadows and around the trunks of trees and in rocky places. If you can get your fingers past the spines, the fruit is delicious when ripe. It's worth husking at least one for a quick sample. Harvesting enough for a pie is another thing all together. I should know! My grandmother sent me out on gooseberry picking expeditions all the time when I was a kid—I had the bloody fingers to prove it. Was it worth it? Yep! The golden brown crust and steaming, sweet filling was outstanding! Have I made one since my grandmother passed? Nope! I'm waiting for one of my grandkids to pick them for me. I shot the flowers along the North Fork of Coffee Creek in late May and the fruit in Cabin Meadow below Sugarpine Lake in September, both in the Trinity Alps.

BALDHIP ROSE

Rosa gymnocarpa — 5 symmetrical petals - ovary superior

Rose family

Flowers of the baldhip rose are not as large as those of the sweet-briar rose (below), but they are every bit as beautiful. They are 1" to 1½" across and a deep magenta or pink. The distinguishing feature of this rose is that the sepals tend to fall off as the hip (seedpod) develops, making the fruit look much like a berry and leading to the name "baldhip" rose. I've found that it tends to grow in shaded or semi-shaded areas in wooded areas at mid-elevations. Good places to see these are along the Canyon Creek Road and the trail to Duck Lakes in the Russian Wilderness. I photographed the flowers along Canyon Creek Road in June.

SWEET-BRIAR

Rosa rubiginosa — 5 symmetrical petals - ovary superior

Rose family

This non-native rose forms large thickets in open and sunny foothill and valley locations throughout the area covered by this book. Individauls form large mounds 6' or more in height and several feet across. Clusters of large, beautiful pink flowers up to 2½" across cover the bushes and are very fragrant! My grandmother loved roses and dug up and transplanted dozens of these to places around her house in Lewiston where, 60 some odd year later, they still thrive.

MOUNTAIN PINK CURRANT, SIERRA CURRANT
Ribes nevadense
Gooseberry family

SIERRA GOOSEBERRY
Ribes roezlii
Gooseberry family

BALDHIP ROSE
Rosa gymnocarpa
Rose family

SWEET-BRIAR
Rosa rubiginosa
Rose family

DOUGLAS SPIRAEA, HARDHACK
Spiraea douglasii 5 symmetrical petals - ovary superior
Rose family

The dark to light rose-pink flowers of Douglas spiraea bloom in narrow, elongated clusters on tall, upright stems from a loosely branched shrub. Prominent stamens give the flower clusters a decidedly fuzzy look. This is a shrub that likes semi-shade and moisture so look for it in damp meadows, around springs and along the banks of creeks in places like Big Boulder Lake and Canyon Creek. **Note:** Intrigued by strange common names, I've always wondered where the term "hardhack" came from. It seems that early settlers in the Northwest gave it that name because it was found in dense stands or clusters that were hard to hack or cut their way through while clearing land or, I suppose, while travelling. Also, it is interesting to note that the scientific name *Spiraea* comes from the Greek word for wreaths or garlands for which reason species of spiraea are often called bridal-wreath shrubs.

ROSE MEADOWSWEET, MOUNTAIN SPIRAEA
Spiraea splendens 5 symmetrical petals - ovary superior
Rose family

The fuzzy looking flat-topped clusters of deep pink flowers on this shrub give this one away. The flowers bloom at the tops of tall stems that can reach to 3′ or so in height. The leaves are oval and scalloped, densley arrayed on the bottom halves of the plant, and only sparsely up the flowering stalk. This spiraea likes semi-shade and moisture so look for it in damp meadows, around springs, and along the banks of creeks and lakes. You will find it at higher elevations in places along Canyon Creek, Long Canyon, and Sugarpine Lake in the Trinity Alps. I shot the large photograph in a wet meadow near Gumboot Lake on the Trinity Divide in July and the small photograph near the outlet of Caribou Lake in September. I have found it blooming around Mirror Lake and in the meadows above Sapphire Lake in the Trinity Alps in September.

WESTERN REDBUD
Cercis occidentalis 5 irregular petals
Pea family

These beautiful shrubs can actually grow into small trees nearly 20′ in height although, locally, 12′ seems to be the norm. They bloom in early spring on bare twigs and are covered from top to bottom with delightful pink pea-shaped flowers which begin to fade as the leaves open. Being in the pea family, the seeds mature in long, reddish-brown pea-like pods. The leaves are mostly round and silver dollar sized or larger. You will see them along almost every road in this neck of the woods. They are a popular spring-time attraction but as pretty as the springtime show is, the colorful autumn show can be just as stunning. When weather conditions are just right the leaves turn beautiful shades of gold and deep red as shown in the small photograph.

DOUGLAS SPIRAEA
Spiraea douglasii
Rose family

ROSE MEADOWSWEET
MOUNTAIN SPIRAEA
Spiraea splendens
Rose family

WESTERN REDBUD
Cercis occidentalis
Pea family

CREEPING SNOWBERRY & COMMON SNOWBERRY
Symphoricarpos albus var. *laevigatus* (left)
Symphoricarpos mollis (right) 5 irregular petals forming a tube
Honeysuckle family

What most people immediately recognize about both species of snowberry are not the flowers but the white berries that mature late in the summer and fall. The inconspicuous, reddish-pink flowers begin to show in early May and are hidden away beneath the leaves making them almost invisible. Not so for the berries which are highly visible. The berries, while inviting, are also poisonous. **DO NOT EAT** and watch young children (who are notorious for putting things in their mouths) around them.

Common snowberry (*S. albus* var. *laevigatus)* is a somewhat stout, erect shrub frequently found in forested areas in deep shade at lower elevations and can reach to four feet or more. It can be found in abundance in places like Clear Creek Campground, along East Weaver Creek, and in abundance along the Trinity River.

The woody, vine-like stems of creeping snowberry (*S. mollis)* are typically reddish brown, grow low to the ground or up into surrounding brush, and can cover large shaded areas from the foothills to the mountains. I've found it along trails above Shasta Lake, along mid-elevation trails in the Trinity Alps and along Trinity Dam Boulevard near Trinity Lake.

I photographed the creeping snowberry flowers above the East Weaver Creek Campground near Weaverville in June and the berries along the trail to Stoddard Lake in the Trinity Alps in September. I photographed the common snowberry fruit (right) in October in Clear Creek Campground on the Trinity Divide.

ROUNDLEAVED SNOWBERRY, MOUNTAIN SNOWBERRY
Symphoricarpos rotundifolius var. *rotundifolius* 5 irregular petals forming a tube
Honeysuckle family

I found this pretty honeysuckle near the outlet of Upper Wright Lake in the Marble Mountains in late July. If I hadn't been wading through neck deep brush. When first placing it according to color in this guide I had some reservations about it belonging in the red to purple section because most of the flowers I found were creamy white and only tinged with light pink. Other photographs online show them as predominantly pink or light magenta. Color aside, the petals are fleshy and joined near the tips which are gently flared and surround several stamens and yellow anthers. The enlarged ovary is inferior with pointed sepals that cup the flowers. Leaves are opposite, round to almost spoon shaped, and scattered along spindly stems which are reddish-brown on the older ones and more greenish on the younger ones. Berries late in the season are white.

COMMON SNOWBERRY
Symphoricarpos albus var. *laevigatus*
Honeysuckle family

CREEPING SNOWBERRY
Symphoricarpos mollis
Honeysuckle family

ROUNDLEAVED SNOWBERRY
Symphoricarpos rotundifolius var. *rotundifolius*
Honeysuckle family

PINK HONEYSUCKLE, HAIRY HONEYSUCKLE
Lonicera hispidula
Honeysuckle family

5 irregular petals forming a tube

As I mentioned in the introductory pages classifying a wildflower by color is sometimes difficult and this honeysuckle is one of those. The flowers emerge in whorled clusters along the stalks and, before they open, appear creamy white with just a hint of pink. When the flowers open and the petals unfurl they fold back upon themselves exposing a beautiful light pink to striking magenta. The opening into the tube is greenish-white with numerous yellowish-green stamens topped with buttery- yellow anthers. The stigma, longer than the stamens, is also pronounced. One of the best indicators that you are indeed looking at a honeysuckle are the opposite, oval leaves which at the ends of the stems become fused, appearing to wrap around the long, trailing stems. Bright red berries ripen later in the summer. This honeysuckle prefers dry, shaded sites. I photographed these along Big French Creek in the Trinity Alps in June. They are also common along lower elevation trails around Shasta Lake.

SWAMP LAUREL
Kalmia polifolia
Heath family

5 petals joined into a bell

Kalmia Lake near the head of Canyon Creek in the Trinity Alps takes its name from this beautiful shrub and for good reason. Like so many alpine lakes in this area, its shoreline is decorated with stands of it and, during the long summer blooming season, it creates a gorgeous setting. The flowers are a striking pink, deeper in hue at the ends of the petals, and almost white in the centers. Each lightly cupped flower has a prominent deep magenta pistil and ten white stamens, with anthers aging blue. It grows to about 4'. Look for it in places like Little Duck Lake in the Russian Wilderness and Virginia Lake in the Trinity Alps. It has a long blooming season, often extending well into September—although, by that time, the bunches of flowers can be pretty spotty. I have found it still blooming in a small, protected pocket along the shoreline of Little Bear Lake in mid-October.

Kalmia Lake in the Trinity Alps takes its name from the beautiful stands of Swamp laurel that fringe its shoreline. This small body of water is set picturesquely with a backdrop of Thompson Peak and Wedding Cake rising beyond the headwaters of Canyon Creek. Add the beautiful blossoms of swamp laurel to the picture and it turns from picturesque to absolutely stunning.

PINK HONEYSUCKLE
Lonicera hispidula
Honeysuckle family

SWAMP LAUREL
Kalmia polifolia
Heath family

Kalmia Lake in the Trinity Alps

PINK MOUNTAIN HEATHER
Phyllodoce empetriformis — 5 petals joined into a bell
Heath family

Worldwide, this is the quintessential wildflower of the high mountains (Scottish Highlanders are especially enamoured with them). It has inspired songs and poetry like no other (except maybe the rose) down through the centuries and with good reason. When it is in its full flower in mid-summer, entire slopes and the shorelines of lakes are alive with its color. It is a shrubby plant that grows to about 4' tall, with lance shaped leaves that appear like bristles along the entire reach of the stems. Beautiful bunches of small, pink, urn shaped flowers hang from the stem ends in thick bundles. You will find them throughout the high country growing at the edges of meadows or next to rocky outcrops. Excellent places to see them are around Caribou Lake in the Trinity Alps and Little Duck Lake in the Russians. In the fall the flowers tend to remain on the stems, turning a pretty golden brown and continuing to put on a nice display long past the flowering season. They are especially nice when the sun backlights them and they literally glow. I shot the large photograph at the back of Big Duck Lake in the Russians and the late season flowers at Little Bear Lake in the Trinity Alps in early October.

DWARF BILBERRY
Vaccinium cespitosum — 5 petals joined into a bell
Heath family

This delightful little huckleberry/bilberry is a low, ankle high shrub which forms dense thickets in shaded, semi-damp areas throughout the Trinity Alps, the Trinity Divide, and Russian Wilderness. It rarely reaches 20 inches in height, has tiny leaves, and greenish stems. The berries are tiny but tasty—if you can find them. Like the berries of its cousin the thin-leaf huckleberry (pgs 112-113), they are much sought after by all sorts of wildlife and humans alike. I photographed the flowers along the Bear Basin Trail in late August and the developing fruit in the meadows at the head of Horse Creek, both in the Trinity Alps. The inlet side of Eaton Lake in the Russians is carpeted ith them.

WESTERN EUPATORY, WESTERN SNAKEROOT
Ageratina occidentalis — Flowers in composite heads
Sunflower family

Look for this beautiful subshrub along streambanks, the shorelines of lakes, and around springs at mid to high elevations throughout this area. It can reach heights of 4' but is more commonly a couple of feet tall with strongly erect stems arising from a woody base. The stems are covered with fine hairs. Leaves are arrowhead shaped, finely toothed, and alternate along the stems. The flowers occur in elongated clusters, can be white, pink, or light purple, and are characterized by prominent style branches, which give the clusters a fuzzy look. I've only found light pink or light purple forms in the high country—never the white version. I photographed these at Big Boulder Lake in the Trinity Alps in late July.

PINK MOUNTAIN HEATHER
Phyllodoce empetriformis
Heath family

DWARF BILBERRY
Vaccinium cespitosum
Heath family

WESTERN EUPATORY
Ageratina occidentalis
Sunflower family

CALIFORNIA SPICEBUSH, CALIFORNIA SWEET-SHRUB

Calycanthus occidentalis

Sweet-shrub family

Many petals

This shrub is a long-time favorite of mine. When it blooms in late spring it is often covered with large, beautiful, deep burgundy flowers. From a few feet away the flowers look velvety soft, but in actuality, are quite rigid. It is common for the sharply pointed petals to be capped with brown at the tips. The leggy shrub grow quickly to heights of 12′ or more and can form dense thickets around springs and along streams throughout the lower elevations of these mountains. The large, deep green leaves are more-or-less elliptic with a rounded base, a pointed tip and are slightly hairy beneath. The bark has a strong smell of camphor if the stems are scraped. I photographed the flower along VanNess Creek above Trinity Lake in June and the seed pod near Clear Creek Campground in October.

Castle Crags Wilderness and State Park: Located on the eastern edge of the Trinity Divide, this spectacular granitic formation rises like ancient battlements above the surrounding wooded hills. The jagged peaks are enjoyed from a distance by travelers along Interstate 5 and by anyone willing to walk the steep trail to the flat below Castle Dome. Visitors include people trekking the Pacific Crest Trail, rock climbers, and day hikers. Wildflowers abound throughout the area with one of the most interesting being the Castle Crags harebell.

CALIFORNIA SPICEBUSH, CALIFORNIA SWEET-SHRUB
Calycanthus occidentalis
Sweet shrub family

Castle Crags in the Castle Crags Wilderness

Masses of Roezl's mountain penstemon grace the edges of Parker Meadows in the Trinity Alps

Silver bush lupine - *Lupinus albifrons* - in bloom above the Trinity River

CROWN BRODIAEA, EARLY HARVEST BRODIAEA, HARVEST BRODIAEA GARLAND BRODIAEA

Brodiaea coronaria — 3 or 6 petals

Brodiaea family

This is another brodiaea that favors dry, open places. Except for height, it and another local species (dwarf brodiaea - *Brodiaea minor*), are similar in basic structure. The deep bluish purple flowers of crown brodiaea, however, tend to be more upright and funnel shaped with petals that are recurved and spread away from the center. Staminodes and stamens are centered and showy. I photographed this bunch on a dry flat along the upper Trinity River in June. Other occurrances are along Stewart's Fork near Trinity Alps Resort. Alice Jones reports it from several areas in Trinity County, around Weaverville for example, and also near Trinity Center in Norwegian Meadows (private property).

BLUE DICKS, WILD HYACINTH

Dipterostemon capitatum — 3 or 6 petals

Brodiaea family

Blue dicks are perhaps the earliest of the brodiaeas to bloom in this area, often appearing on sun-warmed hillsides and exposed roadside cut-banks in early February. They are characterized by the long flowering stalks that are often 18″ or more in height. Each stalk is topped with a tight group of upright, bell-shaped, light to dark blue (sometimes even light violet) flowers. The weight of the flowers at the tops of those tall, slender stems can cause them to bend or undulate, sometimes clear to the ground. I photographed this one on a rocky flat above Clear Creek in Whiskeytown National Recreation Area but it is common at lower elevations throughout this area.

FORK-TOOTHED OOKOW, OOKOW

Dichelostemma congestum — 3 or 6 petals

Brodiaea family

The flowers of fork-toothed ookow bloom in compact heads at the tops of naked stems that can reach to 3′ in height. Individual flowers are about 3/4″ long, range in color from deep blue to purple, are bell shaped, pinched in the middle above the ovaries, and then flaring away at the tips. Each flower has three fertile stamens and three forked, tooth shaped projections in the center surrounding the stamens. The leaves are grooved down the center. I photographed these in a meadow below Rush Creek Road in mid-May. It is found commonly throughout these mountains.

MANY FLOWERED BRODIAEA, ROUNDTOOTH SNAKELILY, WILD HYACINTH

Dichelostemma multiflorum — 3 or 6 petals

Brodiaea family

Brodiaeas are among the earliest wildflowers to bloom in the lower elevations throughout the area covered by this book and this one is no exception. It decorates hillsides from the Central Valley to the hills surrounding the Trinity River and the Trinity Alps. It has compact heads of beautiful light blue to light lilac flowers nodding at the tops of stems that reach 2′ in height. Around the Central Valley their nectar is a favored food source for pipevine swallowtail butterflies. The leaves are grass-like and almost the same length as the flowering stems. I photographed these along the Trinity River near Lewiston but have also shot them above Shasta Lake.

CROWN BRODIAEA
Brodiaea coronaria
Brodiaea family

BLUE DICKS
Dipterostemon capitatum
Brodiaea family

FORK-TOOTHED OOKOW
Dichelostemma congestum
Brodiaea family

MANY FLOWERED BRODIAEA
Dichelostemma multiflorum
Brodiaea family

NAKED MARIPOSA LILY, NAKED STAR TULIP

Calochortus nudus 3 or 6 petals
Lily family

Here is a pretty little lily with an uncertain color palette—the sort of thing that causes amateurs no end of consternation. I have three basic color combinations for this species in my collection ranging from deep purplish-blue to pale pink—all of which are variable. Some of the flower colors are subdued while others are vibrant. Flower structure is consistent however. All have three have petals with ruffled edges fading to near white at the center. Each petal has a vibrant band of deep maroon surrounding the prominent ovary, stamens, and mostly blue anthers. Occassionally that band is missing all together or very faint while in others it extends all the way to the ovary. Small, scattered whitish hairs grow outward from that band. The flowering stalk can host single or multiple flowers along its length. Leaves are narrow, mostly upright, and basal. They are found at mid-elevations in grassy meadows throughout the Trinity Alps, the Trinity Divide, and Russians. Look for them in places like Bowerman Meadows (where I photographed these) and along the Pacific Crest Trail above Kangaroo Lake on the Trinity Divide or in the meadow surrounding Taylor Lake in the Russians. **Note:** This lily is much favored by goldenrod spiders, who frequently take protective custody of the flowers just as they begin to open. They guard their little homes with a vengeance, rushing to the edges of the petals with little legs waving a warning to anyone or anything intruding too closely. I've actually had them attack my camera lens. Go figure! Goldenrod spiders are variable with regards to color—they are yellow, white, or brown.

NAKED MARIPOSA LILY, NAKED STAR TULIP
Calochortus nudus
Lily family

ELEGANT STAR TULIP, NORTHWESTERN MARIPOSA LILY, PUSSY EARS

Calochortus elegans — 3 or 6 petals

HAIRY STAR TULIP, TOLMIE STAR TULIP, WHITE PUSSY EARS

Calochortus tolmiei — 3 or 6 petals

Lily family

Along with millions of other wildflower enthusiasts, I have always known both of these species simply as pussy ears or cat's ears. However, numerous common names exist for these two beautiful species.

These might well be regarded as the true harbingers of spring in the lower elevations of these mountains. There are other wildflowers that bloom earlier but not quite the way these do. Huge areas in open woods and hillsides will be covered with these gorgeous little fuzzy flowers in late March and April. I've found them blooming near the summit of Scott Mountain in early July.

Because both species are variable in appearance, I've included the more common color variations for both . The flowers range in color from white to purplish-blue with combinations in between. But for positive identification purposes It is important to note that they vary in the degree to which the fine hairs cover the top and sides of the petals. *C. tolmie*i has fine hairs that don't extend to the tips of the petals while *C. elegans* has fine hairs that extend to and cover the tips of the petals.

The flowering stalk can grow to around 12" or so and can host several flowers. A single, long, and arching lance-shaped leaf precedes the bloom often followed by several leaves with the flowering stalk or stalks emerging from the axils along the stalk. The photographs of *C. elegans* were taken along Highway 3 near Covington's Mill and on the Swift Creek Trail in the Trinity Alps, in late June and early July. The photograph of *C. tolmiei* was taken in a meadow below Doe Lake in August in the Trinity Alps.

This photograph illustrates the leaf, and structure of the flowering stem common to both species. Remember - no hair on the petal tip - *C. tolmei*, hair on the petal tip - *C. elegans*

ELEGANT STAR TULIP, PUSSY EARS
Calochortus elegans
Lily family

HAIRY STAR TULIP, PUSSY EARS
Calochortus tolmiei
Lily family

BOWL-TUBED IRIS, GROUND IRIS

Iris macrosiphon 3 or 6 petals

Iris family

Members of the Iris family are a beautiful addition to late spring and early summer hikes. But, like most irises, this one is variable in terms of color. It can range from cream to yellow to a deep blue-purple. The key characteristic is a bowl shaped bulge at the top of the long corolla tube. It also hugs the ground and is only about 6″ tall. It is common throughout the lower to mid-elevations in this area. I shot this photograph along the North Fork of the Trinity River Trail in June and have seen it around Natural Bridge in Trinity County just off Highway 3 wouth of Hayfork.

ROCKY MOUNTAIN IRIS, WESTERN BLUE FLAG

Iris missouriensis 3 or 6 petals

Iris family

Unlike the other iris species found in these mountains, this one is tall, often reaching to 36″ or so. It's a gregarious wildflower spreading, like all irises, from a thickened underground rhizome and populations cover large areas. Unlike our other iris species, this one usually occurs in wetlands. The flowering stalk is stout with a single flower blooming at the top. The flower standards (petals) range from light blue to purple, with white falls (sepals) that are yellow near the throat then streaked with blue or purple. Leaves are strongly upright, basal, and lance shaped. It is a beautiful and graceful wildflower and a real treat to photograph. I've seen it above Mill Creek Lake, and along the trail to East Boulder Lake, both in the Trinity Alps, which is where I photographed these. Because of their size and color, they can't be confused with any other native iris in the area covered by this guide.

BLUE-EYED GRASS, IDAHO BLUE-EYED GRASS

Sisyrinchium bellum and *Sisyrinchium idahoense* 3 or 6 petals

Iris family

I've photographed blue-eyed grass from the Trinity Alps to South Carolina where locals call them meadow stars. It is a common wildflower but just inconspicuous enough to easily pass it by. The flowers bloom in clusters at the tops of flattened, grass-like stems, are a light blue with deep blue veining, with a yellow spot at the base, and a sharp pointed tip. *S. idahoense* grows in moist areas around meadows, streams and springs from mid- to high-elevations while *S. bellum* grows at lower elevations. Both species vary in color from light to dark blue to pale lilac. I photographed *S. bellum* on Musser Hill near Weaverville and the inset of *S. idahoense* (inset) in Mumford Meadows along Swift Creek in the Trinity Alps. **Note:** The Latin word *sus* (meaning pig) and the Greek word *rhynchos* (meaning snout) combined allude to pigs digging roots of some bulbous plant for food which is spoken of in the writings of Pliny. To our knowledge the reason for applying the name to a genus of New World irises was apparently arbitrary.

YELLOW TRITELEIA, TRINITY MOUNTAINS PRETTY FACE

Triteleia crocea subsp. *modesta* 3 or 6 petals

Brodiaea family

It's not yellow you say? Here is another of those botanical color anomalies. With its clusters of deep blue and soft purple blooms it only resembles its yellow namesake in structure and habitat. Its geographic distribution is entirely restricted to the Trinity Ultramafic Sheet, which that encompasses the Red Trinities, Scott Mountain, and the Eddys. The stems are strongly erect, to 2′ in height, supporting terminal clusters of those gorgeous flowers. I've found it in scattered populations exclusively in serpentine soils along the road to the North Fork of Swift Creek Trailhead in late May and near Eagle Creek Campground on the Upper Trinity River in mid-June. See page 178 for the yellow version.

BOWL-TUBED IRIS
Iris macrosiphon
Iris family

ROCKY MOUNTAIN IRIS
Iris missouriensis
Iris family

BLUE-EYED GRASS, IDAHO BLUE-EYED GRASS
Sisyrinchium bellum and *Sisyrinchium idahoense*
Iris family

YELLOW TRITELEIA, TRINITY MTNS. PRETTYFACE
Triteleia crocea subsp. *modesta*
Brodiaea family

BREWER'S ROCK CRESS
Boechera breweri 4 petals
Mustard family

You can find this tiny cress in open, dry, rocky areas. It is a small and somewhat inconspicuous wildflower only reaching to about 8″ in height. Lance shaped leaves are scattered the length of the hairy stem. The tube shaped flowers are small, have four deep purple petals, and flare widely at the tips. Flowers rarely exceed 3/8″ in diameter. I photographed these on a rocky outcrop just above Stewart's Fork. You will also find them blooming in serpentine soils just below Parker Divide in the Trinity Alps and in the granite outcrops around Big Blue in the Russians.

WHITESTEM FRASERA, GREEN-GENTIAN
Frasera albicaulis subsp. *nitida* 4 petals
Gentian family

This is a common wildflower from lower elevations in the foothills along the Sacramento River to the high meadows below Red Mountain in the Trinity Alps. Regardless of the elevation, it seems to bloom everywhere in June. It is readily identifiable by its long, grass-like, opposite leaves rimmed with silvery-white. The flowers have four bluish-white petals, are striped, spotted with purple, and each has a small green gland at the base. The petals are pointed and slightly cupped to the margins. The stamens are light lilac and the anthers creamy-white. From a standing distance the flowers don't look like much but get up close and they are striking. I photographed these in an open and brushy area along the upper reaches of the Stoney Ridge Trail in the Trinity Alps.

EXPLORER'S GENTIAN, BLUE GENTIAN, MOUNTAIN BOG GENTIAN
Gentiana calycosa 5 petals
Gentian family

If there is a wildflower more symbolic of the high country than explorer's gentian I don't know what it is. In some circles it is revered as the harbinger of epic hikes to come. The deep, electric blue flowers grow in large colonies and grace mid- to high-elevation trailsides and meadows throughout the area. Their large, deep, urn-shaped blooms are easily recognized. There is one area in the Trinity Alps where I have found, mixed in with the blue, a persistent population of a deep and vibrant pink version. I'm keeping it to myself but feel free to go looking! If you find them let me know. I've photographed these in several areas including Upper Wright Lake in the Marble Mountains, at Grizzly Lake in the Trinity Alps, and elsewhere.

FRINGED GENTIAN, HIKER'S GENTIAN, ONE-FLOWERED FRINGED GENTIAN
Gentianopsis simplex 4 petals
Gentian family

This is a small, upright gentian common in wet meadows throughout the Trinity Alps. It rarely exceeds 8″ in height and hides in the surrounding plants—making it difficult to see. A solitary tube shaped flower blooms atop a mostly naked stalk with petals that, as they emerge from the calyx, are creamy white mottled with blue on the exposed side. The showy part of the petals are a rich slate blue with margins that are ruffled and slightly toothed or "fringed" as its common name suggests. Spoon shaped leaves are basal but small linear leaves do occur on the lower parts of the flowering stalk—both are often dried by the time the flowers appear. I photographed these in the meadows below Trail Gulch Lake in the Trinity Alps in July. Years ago I found a population of white ones interspersed with the blue along the old trail into Stoddard Lake.

BREWER'S ROCK CRESS
Boechera breweri
Mustard family

WHITESTEM FRASERA - GREEN-GENTIAN
Frasera albicaulis subsp. *nitida*
Gentian family

EXPLORER'S GENTIAN, BLUE GENTIAN
Gentiana calycosa
Gentian family

FRINGED GENTIAN, HIKER'S GENTIAN
Gentianopsis simplex
Gentian family

DAGGERPOD, PHOENICAULIS, WALLFLOWER PHOENICAULIS
Phoenicaulis cheiranthoides 4 petals
Mustard family

On Packer's Peak (7,828') in the Trinity Alps, on a cold, windy day in May with snow and rain showers moving across the high country, I found a small population (three plants to be precise) of this gorgeous wildflower blooming in protected rocky crevices. Even though I did a wide search around the summit I found no others. This native plant is common throughout the Pacific Northwest and is not picky as to habitat. It is found on volcanic and sandy soils, in rocky areas, and in meadows and sagebrush scrub from around 3,200' to 10,500'. The light green and fuzzy leaves form a small and mostly upright basal rosette. Leaves are more or less lance shaped and occur alternately along the flowering stem. Beautiful purple flowers cluster at the tops of the flowering stalks. Even though it is common, I have to admit this was the first time I'd ever encountered it and I've been to the top of Packer's Peak many times.

STAR SWERTIA
Swertia perennis 5 petals
Gentian family

This pretty little wildflower is an anomaly in this area—with only a few small, isolated populations on the Trinity Divide and Scott Mountains. This is far removed from other known California populations in the Central and Southern Sierra. It's a worldly flower, however, found in the Wallowa Mountains of Oregon, the Rocky Mountains, Alaska, and Eurasia. In 2018 I photographed it in Switzerland where it is called Moorenzian or marsh felwort. Multiple flowers bloom along a fairly tall stalk that reach about 14". Leaves are basal and somewhat spoon-shaped or obovate. The deep violet-blue flowers are streaked with a silvery-blue and each petal has two small, deep, purple fringed ridges located between the stamens as illustrated in the inset photograph. Stamens are prominent, deep purple, and topped with purple anthers. These wildflowers favor wet meadows and bogs. Almost all of those I photographed on this particular day were just opening in mid-September

SERPENTINE SNOW QUEEN 4 petals
Synthyris cordata
Plantain family

Step in an unseen hole on a steep hillside, fall down, and while recovering and wondering what happened, you discover an entirely new-to-you wildflower. This "fortunate accident" actually happened to me and I believe that nifty discoveries are often made at moments least expected. These absolutely beautiful, soft blue, tubular to cup shaped flowers have petals that unfurl with two large and strikingly pretty rose-pink anthers protruding well beyond the petal tips. The flowers bloom in clusters at the ends of short reddish-green, slightly hairy stalks. The deep green leaves are arrowhead shaped and hairy with cleft and wavy margins. All in all, this is a most beautiful plant and I count myself lucky for having suffered nothing more than a skinned elbow during this discovery. I shot these photographs above Burnt Ranch Falls on the Trinity River in early February.

DAGGERPOD, PHOENICAULIS
Phoenicaulis cheiranthoides
Mustard family

STAR SWERTIA
Swertia perennis
Gentian family

SERPENTINE SNOW QUEEN
Synthyris cordata
Plantain family

MONKSHOOD
Aconitum columbianum — 5 irregular petals
Buttercup family

Monkshood is common throughout this area and favor wet meadows at mid-elevations. The flowers are a deep blue or purple but it is not uncommon to find creamy-white ones as well. The distinctive hood shape is a dead give-away and it cannot be mistaken for any other flower. Typically they range to about 16" in height but I have seen them to around 4' . I photographed the purple ones in Long Canyon and the white one near the inlet to Rush Creek Lake (on the north side of the Trinity Alps).

WOOLY-POD MILKVETCH, PURSH'S MILKVETCH
Astragalus purshii — 5 irregular petals
Pea family

This tiny vetch is as pretty when it goes to seed as it is when it's in full bloom. It is so small that I've seen people walk right over whole patches of it without realizing it was there. It grows on open and exposed serpentine flats and slopes and seems especially abundant on Scott Mountain. The deep pink-purple flowers are about ½" long and the fuzzy, globose seed pods are its most prominent features. The pinnately compound leaves are gray-green, fuzzy, and grow low to the ground. Step carefully, go slowly, and keep your eyes open—it's the only way you will ever see them. I shot the photograph of the flower along the road to Big Carmen Lake on Scott Mountain in early June and the seeds in September.

GIANT LARKSPUR, TALL LARKSPUR, MOUNTAIN LARKSPUR
Delphinium glaucum — 5 irregular petals
Buttercup family

This larkspur is easily identifiable by sheer size alone. Plants can reach 8' in height with upper stems covered in deep blue to purple flowers. It is unmistakable. The stems and deeply divided leaves are pale green. It favors wet meadows or proximity to running water. I photographed these in Bear Basin in the Trinity Alps but they are a standout throughout the area covered by this guide. The tallest I have ever seen grow in the meadows below Buck Lake in the Trinity Alps and in the meadows surrounding Deep Lake in the Marble Mountain Wilderness.

DWARF LARKSPUR, NUTTALL'S LARKSPUR
Delphinium nuttallianum — 5 irregular petals
Buttercup family

This is another of the earliest bloomers at lower elevations and can be found almost everywhere from the foothills to mid-elevation meadows in the high country. Plants grow to 16" tall with rounded leaves that are palmately divided. Several deep blue to purple flowers bloom along a strongly erect stem. The petal-like sepals fan away from the center and the upper petal has a toothed and white decoration that forms a "hood" which arches over the stamens and pistils. The spur at the back of the flower is about ½" long. I photographed these along Canyon Creek in late May but look for them along almost any trail in the area. **Note:** Goldenrod spiders are pretty fond of this wildflower so be on the lookout for them! I photographed this one in Union Meadows.

MONKSHOOD, COLUMBIA MONKSHOOD
Aconitum columbianum
Buttercup family

WOOLY-POD MILKVETCH
Astragalus purshii
Pea family

GIANT LARKSPUR, TALL LARKSPUR
Delphinium glaucum
Buttercup famil

DWARF LARKSPUR, NUTTALL'S LARKSPUR
Delphinium nuttallianum
Buttercup family

LEAFY PEA, OREGON PEA 5 irregular petals
Lathyrus polyphyllus
Pea family

Walking out of the East Boulder Lake Basin, in a small meadow bounding the trail on either side, I noticed the deep purple blooms of this little pea. I dropped my pack, dug out my tripod and camera, and set about taking some photographs. Closer inspection of the surroundings revealed that the "meadow" was actually made up almost exclusively of this pea interspersed and a small population of thymeleaf speedwell. The younger flowers, not yet fully opened, were a beautiful deep magenta-purple emerging from hairy reddish sepals. Newly-emerging flowers were tinged with light pink fading to white just below the sepals and turning bright magenta closer to the outside petal tips. The banner and keel were bluish purple streaked with veins of deep purple. The undersides of the keels were whitish tinged with light purple. This lush little pocket meadow was shaded and the soils damp beneath a canopy of mixed conifers. I shot this photograph in mid-June.

Note about lupines: As mentioned at the beginning of this guide, certain wildflowers are difficult to properly identify— especially small to medium sized lupines. The upshot is that if the lupine is similar to one of those shown in this guide you can be reasonably sure that you've made a good ID. That being said, don't be surprised if you just can't make the real thing jive with the photograph presented here. For most of us just calling it a lupine will be sufficient.

SILVER BUSH LUPINE
Lupinus albifrons 5 irregular petals
Pea family

This wildflower is a definite winner in the roadside attraction category. It is found throughout the area and because of its size (woody stems up to 3′ tall), the masses of deep purple flowers, and gorgeous silvery foliage, it is unmistakable. Take a drive almost anywhere and you will see them. It is one of my favorite early bloomers and one that does well in native plant gardens. A great place to enjoy them is at the Whiskeytown National Recreation Area Visitor Center. I photographed these along the Trinity River just below Lewiston.

MINIATURE LUPINE, TWO-COLORED LUPINE
Lupinus bicolor 5 irregular petals
Pea family

This diminutive lupine is an early bloomer, often found at the beginning of May on grassy flats and hillsides in lower elevations. It often escapes notice because it rarely grows over 6" in height. Leaves are deeply cleft, dark green, and found along the lower portion of the stem. The flowers are deep bluish-purple, with a white patch on the upper flag, decorated with tiny purple spots and bloom scattered along the erect stem. I photographed these near Squirrel Flat on the north end of Trinity Lake in early June.

LEAFY PEA, OREGON PEA
Lathyrus polyphyllus
Pea family

SILVER BUSH LUPINE
Lupinus albifrons
Pea family

MINIATURE LUPINE
TWO-COLORED LUPINE
Lupinus bicolor
Pea family

BREWER'S LUPINE
Lupinus breweri var. *breweri* 5 irregular petals
Pea family

Brewer's lupine is found at higher elevations in the backcountry. It is small and rarely exceeds 4" to 6" in height—including the flowers which bloom in little compact heads, often forming extensive and dense mats. The stout leaves of this lupine are covered with silvery hairs. I have found populations in which the flowers were not much larger than the penny featured in the inset. While hiking along the ridgetop from Upper Albert Lake to Taylor Lake in the Russians, I found populations of them that quite literally covered acres of exposed decomposed granite soils. I found the same thing on the Bear Creek trail between East Weaver Creek and Monument Peak in the Trinity Alps. I photographed this one on the ridge above Twin Lakes in the Russians in late August. A penny next to one of the flowers gives some indication of size.

BROAD-LEAVED LUPINE
Lupinus latifolius 5 irregular petals
Pea family

This is a common trailside lupine that you will find growing throughout the area at low to mid-elevations. The flowering stalks reach to 2' in height. The flowers tend to a light lilac with deep purple accents. Leaves are vibrant green. Overall, this wildflower is closely associated with shaded areas and relatively damp soils. I photographed these along the trail to Big Duck Lake in the Russian Wilderness. I have also photographed them along the lower reaches of the East Fork of Stewart's Fork Trail.

LOBB'S LUPINE
Lupinus lepidus var. *lobbii* 5 irregular petals
Pea family

Lobb's lupine is short-in-stature and forms dense mats on sandy soils at higher elevations. It favors decomposed granite flats and slopes in open sun, at or near the tops of ridges. The light blue to deep purple flowers form compact heads approximately 2" in diameter and 2" tall. Leaves are stiff, deeply cleft, and covered with fine hairs. These could easily be mistaken for Brewer's lupine except these tend to be a bit more sprawling in habit, taller, and the flowers not nearly as compact. I photographed these on the broad ridge between Lost Lake and Cub Wallow in the Trinity Alps in early August where they covered fairly large areas.

VELVET LUPINE
Lupinus leucophyllus 5 irregular petals
Pea family

This is probably the most unique lupine you will ever find in this area. It is also perhaps my favorite. The plant is soft, fuzzy, silvery-white, and tinged with magenta on the stems and leaf petioles. The flower petals are creamy-white to light purple or lavender and age quickly to a pale brown. When I say that the plant is fuzzy and soft I mean kitten soft. It's one of those plants you want to pet! I have photographed these alongside Highway 3 near Stewart's Fork and alongside the road to the Stoney Ridge Trailhead. They don't grow anywhere in abundance but both of these populations include a few dozen specimens. The tallest plants grow to about 24" give or take. To fully appreciate them you need to get up close and personal because, from a distance they are non-descript—but believe me, they are not!

BREWER'S LUPINE
Lupinus breweri var. *breweri*
Pea family

BROAD-LEAVED LUPINE
Lupinus latifolius
Pea family

LOBB'S LUPINE
Lupinus lepidus var. *lobbii*
Pea family

VELVET LUPINE
Lupinus leucophyllus
Pea family

Blue overload: There are lupines and then there are the lupines that can treat your eyes to a true blue magnificence. These two—valley sky lupine and bigleaf lupine—are two such wildflowers. They bloom in such large colonies that they cover large stretches of the landscape and your imagination. They are wildflowers worth celebrating!

VALLEY SKY LUPINE, SKY LUPINE
Lupinus nanus
Pea family

5 irregular petals

This might be the most recognizable lupine anywhere in this area. Beginning in late March it lights vast areas of the landscape in clouds of blue from the floor of the Central Valley to the foothills of the Trinities. No other lupine in this area puts on a show such as this one. The flowers are arranged in concentric whorles around a stalk which range from 8" to 12" in height. Its prefers open and sunny areas that, by July, have turned the classic California brown. A drive along Interstate 5, Highway 299 and 3, or even up Canyon Creek Road will reward you with fields of color. I shot these photographs along Canyon Creek. **Note:** There are several alternative common names for this flower including field lupine, dwarf lupine, ocean-blue lupine (particularly nice I think) and Douglas annual lupine.

BIGLEAF LUPINE
Lupinus polyphyllus
Pea family

5 irregular petals

This is, perhaps, the showiest of all the lupines in the high country—mainly because of its size. The flowers are arrayed in tall spikes with the newly opened blooms at the top a beautiful soft purple fading to near white with the older blooms at the bottom. It is especially abundant in wet meadows in the Russian Wilderness. I photographed these in the meadow at Lower Albert Lake in September where I found hundreds blooming 3' tall stalks. I found them in bloom in the meadow around Deep Lake in the Marble Mountain Wilderness and they have been reported from around Castle Lake on the Trinity divide.

VALLEY SKY LUPINE, SKY LUPINE
Lupinus nanus
Pea family

BIGLEAF LUPINE
Lupinus polyphyllus
Pea family

WESTERN DOG VIOLET, EARLY BLUE VIOLET
Viola adunca subsp. *adunca* — 5 irregular petals
Violet family

These little violets bloom as early as late April in vernally damp places at lower elevations in the high country, but May is a better bet. They are definitely a "get down in the dirt" wildflower if you really want to check them out. The little, elliptic, cup-shaped leaves have crenulate edges and show as the ground warms in the spring. The leaves are followed by flowering stalks about 1" to 1 ½" tall topped with a single pale blue or deep purple flower. I shot this photograph along Union Creek and in Bullard's Basin in the Trinity Alps in early June. **Note:** The leaves and flowers of this violet are high in vitamin A and C, are edible, and used in salads. They can also be brewed as tea. The rest of the plant—roots, fruit, and seeds are poisonous.

LYALL'S ANEMONE
Anemone lyallii — 5 symmetrical petals - ovary superior
Buttercup Family

If you want to see this beautiful little anemone walk the trail to Little Duck Lake in the Russians in early summer—it's the only place in this area I've ever seen it. The little flowers, 1" to 1 ½" in diameter, have five to six sepals that resemble petals and, in this area, range from pale blue to light purple to deep magenta. The predominant color, however, is a pale bluish-purple. Lyall's has fewer than 35 stamens as opposed to more than 35 for Oregon anemone which, in all other aspects, it closely resembles. The flowering stalks are fairly tall with a single large basal leaf that is divided into three leaflets and then divided again. It is often withered by the time the plant flowers. There are three additional whorled 'leaves' along the stem below the flowers, called bracts, which look similar to the basal leaf. It grows in proximity to wood anemone (page 348). I took this photograph on the trail to Little Duck Lake in July.

WESTERN BLUE FLAX
Linum lewisii var. *lewisii* — 5 symmetrical petals - ovary superior
Flax family

This is one of the iconic wildflowers found in the dry, open, upper elevation, high country meadows throughout these mountains. The graceful and large blue flowers bloom loosely at the top of tall, erect stems that range to 14" or more in height. The stems are covered with short, narrow leaves. The broad, dish-shaped flowers are about 1" to 1 ¼" in diameter and have five petals that are striped lengthwise with a deep blue. They never seem to bloom in large populations but they do congregate over large areas. I photographed these in Siligo Meadows in the Trinity Alps in early August.

JESSICA'S STICKSEED, MEADOW STICKSEED
Hackelia micrantha — 5 symmetrical petals forming a tube
Borage family

This is a common wildflower in the mid-elevations. The tiny, pink to light blue flowers bloom in loose bunches scattered along tall stems that can reach 20" in height. They love wet meadows, the banks of streams, and shorelines of lakes and ponds. When the seeds mature in late summer they literally stick to anything and everything—especially socks! I photographed these along the outlet stream below Doe Lake in August but you can find them almost everywhere.

WESTERN DOG VIOLET, EARLY BLUE VIOLET
Viola adunca subsp. *adunca*
Violet family

LYALL'S ANEMONE
Anemone lyallii
Buttercup family

WESTERN BLUE FLAX
Linum lewisii var. *lewisii*
Flax family

JESSICA'S STICKSEED, MEADOW STICKSEED
Hackelia micrantha
Borage family

THE BLUE-EYED MARYS (Collinsias): The four species of this little wildflower that I have photographed in these mountains are respresented below. Another species - *Collinsia parviflora* or few-flowered blue-eyed Mary is regional but not pictured.

LARGE-FLOWERED BLUE-EYED MARY
Collinsia grandiflora
Plantain family

5 irregular petals forming a tube

Large-flowered, in this case, refers to its size relative to the other blue-eyed Marys. They range to 10" in height and are loosely branched with flowers blooming in whorl-like patterns along the upper part of the stalk. The upper two petals of the flowers flare upward and are white with a tinge of purple or blue. The lower petals hang downwards and are dark bluish-purple. Flowers are rarely more than 1/4" in diameter. They like open, sunny areas blending in nicely with their surroundings. I photographed these just above Canyon Creek in early June.

NARROW LEAF COLLINSIA, NARROWLEAF BLUE-EYED MARY
Collinsia linearis
Plantain family

5 irregular petals forming a tube

I've been up and down the Packer's Peak Trail so many times that I would like to think I know everything there is to know about it. But such is not the case. In this instance, I found this pretty little collinsia–one I'd never encountered before–and as usual it was hiding in plain sight. The only difference between this particular trip and those before was the timing–it's always about timing when hunting for wildflowers. This collinsia has narrow, lance shaped leaves that look like fir needles. They are arranged in whorls around the flowering stalk with the flowers emerging from the axils. The top petals are mostly white edged with blue streaks from the center. There are faint creamy yellow spots near the throat. The lower petals are deep blue–almost purple. I shot this photograph in early June.

HILLSIDE COLLINSIA
Collinsia sparsiflora var. *collina*
Plantain family

5 irregular petals forming a tube

This tiny collinsia blooms in huge populations in areas above Shasta Lake and also along the old Delta Toll Road near its junction with Trinity Mountain Road on the Trinity Divide. It doesn't look like much until you get up close and personal with it. There can be over a dozen flowers per plant, borne in the leaf axils on short stalks. They vary from light pink to light blue with upper petals flaring away from the lower ones–each with a hairy basal spur. They are white in the throat becoming blue or pink at the edges with deep maroon spots only on the white part. The basal spur is mostly yellow with maroon spots. Lower petals are mostly pink or blue and droop slightly from the throat. Plants range from 6" or 7" tall. Leaves are narrow and tapered at the tip. I shot these photographs in late March.

TORREY'S BLUE-EYED MARY
Collinsia torreyi var. *latifolia*
Plantain family

5 irregular petals forming a tube

This is one of the tiniest collinsias in this area. Plants are only a few inches tall, loosely branched, with flowers that are often less than 3/16" in diameter. The top two petals are reflexed, white or creamy white, lightly streaked with blue, with purple spots. The lower petals are electric blue tinged with purple. Stems and sepals are hairy and covered with sticky glands. Look closely for it in sandy soils alongside trails throughout these mountains. I have to say, though, that even looking intently for it, you might miss it–it is so difficult to spot. I photographed these above Little Granite Lake on the trail from Doe Lake in late August.

LARGE-FLOWERED BLUE-EYED MARY
Collinsia grandiflora
Plantain family

NARROW LEAF BLUE-EYED MARY
Collinsia linearis
Plantain family

HILLSIDE COLLINSIA
Collinsia sparsiflora var. *collina*
Plantain family

TORREY'S BLUE-EYED MARY
Collinsia torreyi var. *latifolia*
Plantain family

PALE PENNYROYAL, COYOTE-MINT, WESTERN MOUNTAIN BALM, PALE MOUNTAIN MONARDELLA, PALE MOUNTAIN MINT
Monardella odoratissima 5 irregular petals forming a tube
Mint family

So many names to choose from but for me it all boils down to this: I find it to be one of the most delightful plants in this book. You know immediately when you're in the vicinity by the wonderful smell of mint. It grows everywhere from the foothills surrounding Shasta Lake to higher elevations in the back-country. It isn't particularly picky about habitat except that it shies away from wet areas, preferring dry and sunny spots. The plant is loosely structured, about 8" to 10" tall, with stiff stems, and lanceolate leaves that range from deep to light grayish-green. Crush the leaves and the mint smell will stop you in your tracks. Flower heads are flat to slightly globose with dozens of individuals beginning their bloom around the edges and gradually working towards the center. Individual hairy buds and flowers vary from lilac to light magenta to white. All in all, this one is hard to miss. I photographed these below Josephine Lake in the Russians in mid-August.

SISKIYOU BEARDTONGUE
Penstemon anguineus 5 irregular petals forming a tube
Plantain family

This is a common penstemon at lower elevations that blooms in moist, shaded areas throughout the area covered by this book. The plants reach 12" in height with loosely scattered whorls of pale lavender flowers along the top 1/3 of the stem. Flower color is the most telling feature for identification. Stems and leaves are somewhat hairy. I took this photograph along the Bowerman Meadows Trail in the Trinity Alps in July.

AZURE PENSTEMON, BLUE PENSTEMON
Penstemon azureus 5 irregular petals forming a tube
Plantain family

Who doesn't love a penstemon? The showy, electric blue flowers decorate trails and hillsides throughout the Klamath Mountains from early spring into late fall. These flowers are fairly large, with stems up to 30" long—especially at lower elevations. Flowers open from yellowish buds, that develop into deep lavender tubes, further opening into striking azure blue petals. Stems and leaves are hairless and the plant's base is woody. It grows on open, dry slopes. I photographed these on Slate Mountain on the Trinity Divide in May and above Upper Wright Lake in the Marbles in July. Look for them also along the lower stretches of the trail to Duck Lakes in the Russians. They are abundant in the foothills around Redding where flowers can persist well into the hot summer season.

PALE PENNYROYAL COYOTE-MINT
Monardella odoratissima
Mint family

SISKIYOU BEARDTONGUE
Penstemon anguineus
Plantain family

AZURE PENSTEMON
Penstemon azureus
Plantain family

THREAD-LEAVED BEARDTONGUE, THREADLEAF PENSTEMON
Penstemon filiformis — 5 irregular petals forming a tube
Plantain family

Several thin, 10" - 20" flowering stalks arise from the woody rootstock and are surrounded by a thick basal cluster of narrow, thread-like leaves—also found along lower part of the stalks. The leaves are mostly opposite and entire but sometimes finely toothed. Flowers are deep bluish-purple and less than ½" long. It is endemic to the Klamath Mountains and found predominantly on serpentine soils where it prefers the edges of open forested areas or the edges of meadows. I have also found it growing around the periphery of sandy and rocky stream terraces. I photographed these along the Upper Trinity River in early June. **Note:** This species, along with *P. azureus*, *P. heterophyllus*, *P. roezlii* and some others, are part of a small group of beardtongues that have distinctive arrowhead-shaped anthers.

SMALL PENSTEMON, SHORT STALK PENSTEMON
Penstemon parvulus — 5 irregular petals forming a tube
Plantain family

Even though the flowers of this penstemon are small they stand out like neon lights on a dark night—even in the bright light of a noonday sun. The flower stalks are erect and without hairs or glands. The ¾" long, deep blue to pale violet flowers hug the tall, thin, stems. The narrow gray-green leaves are mostly lanceolate but can be narrowly ovate and clasp the stem for over half the length of the flowering stalk. Several flowering stalks arise from a woody base which is where most of the leaves are congregated. I photographed these near Eagle Creek and the Upper Trinity River in early June where it was blooming in open, sandy and rocky stream terraces. Consider your identification of this little penstemon closely because it could be mistaken for thread-leaved penstmon.

WHORLED PENSTEMON, SMALL FLOWERED PENSTEMON
Penstemon procerus var. *brachyanthus* — 5 irregular petals forming a tube
Plantain family

Here's another of my favorites. This beautiful and unique penstemon blooms at mid- to high-elevations in and around wet meadows, damp open hillsides, and the shorelines of lakes. I hardly know of a lake anywhere in the high country that isn't home to a population of them. The loosely crowded whorls of deep lavender to lilac colored flowers adorn the top half of strongly upright, hairless stems. I took this photograph above Horse Creek Meadows in the Trinity Alps in July. They are common around Upper Albert Lake in the Russians as well.

WHORLED PENSTEMON, PINCUSHION PENSTEMON
Penstemon procerus var. *formosus* — 5 irregular petals forming a tube
Plantain family

This pretty little penstemon is one that blooms at higher elevations in these mountains, usually on exposed rocky outcrops and hillsides. The flowers are arranged in tight whorls (on shorter stalks than P. *brachyanthus)* and tend to be light pink fading to light blue but sometimes an overpowering electric pink. One of the most beautiful displays I've ever seen was on the ridge above Papoose Lake in the Trinity Alps where there were hundreds of them all in bloom at the same time and where I shot this photograph. That was in late August which is pretty typical of this high country variety. Other good places to see them are on the rocky ridges above Caribou, Mirror, and Kalmia Lakes, all in the Trinity Alps.

THREAD-LEAVED BEARDTONGUE
Penstemon filiformis
Plantain family

SMALL PENSTEMON, SHORT STALK PENSTEMON
Penstemon parvulus
Plantain family

WHORLED PENSTEMON
Penstemon procerus var. *brachyanthus*
Plantain family

WHORLED PENSTEMON, PINCUSHION PENSTEMON
Penstemon procerus var. *formosus*
Plantain family

ROEZL'S MOUNTAIN PENSTEMON, PURPLE PENSTEMON

Penstemon roezlii 5 irregular petals forming a tube

Plantain family

THis is another wildflower that can cause some confusion when sorting by color. Are they blue, purple, or magenta? For the most part they are a mix of magenta at the throat progressing to electric blue at the petal tips. The lower petals sport two white stripes that emerge from the throat. Flowers are small—about ½″ long—and crowded along one side of a short reddish stalk. The lance shaped leaves are about ½″ long and attached alternately along the lower third of the flowering stalk which, in turn, arises from a woody base. Preferred habitat is relatively dry, exposed serpentine slopes and flats. I've found it blooming on Scott Mountain in June, below Stonewall Pass in July, and above the Upper Trinity River in late May. The large photograph was taken on Scott Mountain and the inset near Eagle Creek. One of the most stunning displays I've ever seen was in Parker meadows in mid-July where there were thousands of them in bloom.

LANCE LEAF SELF-HEAL, MOUNTAIN SELFHEAL

Prunella vulgaris subsp. *lanceolata* 5 irregular petals forming a tube

Mint family

This little member of the mint family likes moist shaded sites in deep woods, wet meadows, streams banks, and springs and seeps. It is common at mid- to high- elevations throughout the area. The two-lipped flowers bloom in densely packed spikes and are either deep blue or pinkish-magenta fading to white at the edges. The upper lip forms a concave cover over the lower lip which hangs mostly straight down. I photographed the blue version along the trail to Stoddard Lake and the pink one near the outlet to Sapphire Lake. Horse Heaven Meadows on the Trinity Divide is another good place to look.

CREEPING SAGE, SONOMA SAGE

Salvia sonomensis 5 irregular petals forming a tube

Mint family

This is a low elevation wildflower, forming large colonies in open, rocky sites. The plants are about 6″ to 8″ tall with stout, strongly erect stems—encircled with whorls of sticky bracts and dozens of light lilac-blue flowers. The lower petal of each flower extends well beyond the others and is cleft and ruffled, with deep purple filaments topped with bright yellow anthers extending well beyond the flower. It is widespread and common throughout these mountains. I shot this photograph alongside the road to the Granite Peak Trailhead on Buck Ridge in May.

ROEZL'S MOUNTAIN PENSTEMON
PURPLE PENSTEMON
Penstemon roezlii
Plantain family

LANCE LEAF SELF HEAL
MOUNTAIN SELFHEAL
Prunella vulgaris subsp. *lanceolata*
Mint family

CREEPING SAGE
Salvia sonomensis
Mint family

NOSE SKULLCAP, SNAPDRAGON SKULLCAP

Scutellaria antirrhinoides

5 irregular petals forming a tube

Mint family

This tiny skullcap is easily missed. They almost always occur as scattered individuals and hardly ever as large populations. The stems are strongly upright, about 6" to 8" tall, with two oval leaves arranged in opposite pairs along the stem. One soft purple flower blooms in each leaf axil, so that the flowers are paired up the stem. They like grassy meadows, open rocky hillsides, and full sun. I shot the large photograph above Canyon Creek on an exposed granite shelf where it was blooming next to California skullcap and the hybrid version *Scutellaria californica X antirrhoides* (page 94). I shot the inset photograph in Mumford Meadows above Swift Creek in the Trinity Alps in late August.

GRAY-LEAVED SKULLCAP

Scutellaria siphocampyloides

5 irregular petals forming a tube

Mint family

This skullcap often reaches 12" in height with narrowly oval, smooth leaves scattered along the stout, upright stem. The purple flowers are borne in the leaf axils and are tubular until the upper 1/3 when the petals begin to flare. The upper petal narrows to form a small rounded protuberance that overarches the lower petal. The lower petal flares outward and has a clear, almost rectangular white spot in the center. I photographed this one in the wet meadows along the Lake Eleanor trail in the Trinity Alps.

DANNY'S SKULLCAP, TUBEROUS SKULLCAP

Scutellaria tuberosa

5 irregular petals forming a tube

Mint family

Like the other skullcaps found in this area, this one is a small plant growing to about 8". The stems that are hairy and red to reddish-green. The leaves are oval and hairy— as are the flowers. Flowers bloom out of the leaf axils and are deep purple to blue. The upper petal forms a hood that overarches the lower petal, which is flared, ruffled, with a white patch filled with purple spots. I photographed this one in the drier part of the meadow surrounding upper Twin Lake above Swift Creek in July. I've not found it elsewhere.

NOSE SKULLCAP
SNAPDRAGON SKULLCAP
Scutellaria antirrhinoides
Mint family

GRAY-LEAVED SKULLCAP
Scutellaria siphocampyloides
Mint family

DANNY'S SKULLCAP
TUBEROUS SKULLCAP
Scutellaria tuberosa
Mint family

WATER SPEEDWELL

Veronica anagallis-aquatica 5 irregular petals forming a tube

Plantain family

This little naturalized veronica is native to Europe but found widely around North and South America. In this area look for it along the banks of the Trinity River where it is quite common and often grows with its feet in the water. It also likes wet meadows. The plants are loosely branched and somewhat spreading. The pale lilac flowers have deep violet or blue streaks eminating from near the throat. The flowers are tiny, no more than 1/4" in diameter. I photographed these in the Trinity River near Junction City in mid-August. I've also seen them along the lower stretches of the North Fork of the Trinity River as well.

AMERICAN SPEEDWELL, AMERICAN BROOKLINE

Veronica americana 5 irregular petals forming a tube

Plantain family

I found and photographed these little beauties in late June while wandering around the meadows below Lower Deadfall Lake on the Trinity Divide. There were several populations of them blooming in and around water in the deep shade provided by stands of alder and willow. Individual plants were small, no more than 8" tall, and loaded with tiny blue flowers. The petals are streaked with deep magenta and white emanating from a green ovary at the center. Stamens are prominent and snow white. Leaves are deep green and slightly oval with toothed edges. According to the website **Raven's Ruff Stuff And Other Things Native** this plant has multiple uses including as a salad green superior to water cress, as a scalp conditioner, and as a treatment for respiratory distress!

COPELAND'S SPEEDWELL

Veronica copelandii 5 irregular petals forming a tube

Plantain family

This pretty little wildflower is endemic to the high Klamath Mountains of Siskiyou and Trinity Counties. The plants are small with erect stems rarely more than 6" tall. The stems are topped with clusters of small flowers. The petals vary from rounded to wedge shaped and are of unequal size—the broadest petal at the top and the most pointed at the bottom. They are light purple near the tips with a white patch at the center surrounded by a deep purple band. The two long stamens are a conspicuous feature, extending well beyond the petals. The stems are reddish and the opposite, oblong, pointed leaves are covered with fine hairs and are sticky. I photographed these in the upper reaches of Long Canyon in July. Another place to see them is along the Pacific Crest Trail and the basins above Kangaroo Lake on the Trinity Divide.

THYMELEAF SPEEDWELL

Veronica serpyllifolia 5 irregular petals foring a tube

Plantain family

I was photographing another wildflower in a trailside meadow when I spotted this tiny veronica a few feet away. A closer look revealed dozens of the little plants with several delicate blue fading to white pea-sized flowers along the 5" to 6" stalk. Each tube, or slightly cup shaped flower, sat atop a longish stalk rising out of the leaf axils. Flowers opening near the tops of the stalks displayed a brilliant sky blue while the older flowers farther down the stalk were a softer blue. I shot these photographs in mid-June near East Boulder Lake.

WATER SPEEDWELL
Veronica anagallis-aquatica
Plantain family

AMERICAN SPEEDWELL
Veronica americana
Plantain family

COPELAND'S SPEEDWELL
Veronica copelandii
Plantain family

THYMELEAF SPEEDWELL
Veronica serpyllifolia
Plantain family

PURPLE MILKWEED
Asclepias cordifolia
Dogbane family

5 symmetrical petals forming a tube

The various species of milkweeds, including this one, are a common sight along the edges of low elevation roads and highways throughout the area. They tend to grow in large colonies and are tall—sometimes approaching 2′ or more. This one is covered with large bunches of beautiful purple and white flowers with the corolla forming a high reflexed arch. The individual flower stems are long, flexible, and the same color as the flowers. The leaves are large and dark green. Break the stem and a sticky, latex type milk appears. I photographed these along Highway 3 near Callahan. **Note**: Butterflies love milkweed. It is an important food source for them

GRAND HOUND'S TONGUE
Adelinia grandis
Borage family

5 symmetrical petals forming a tube

Grand hound's tongue is among the earliest wildflowers to bloom in the foothills. It takes its name from the leaves that look a lot like a hound's tongue—just a little bit more on the green side! Leaves are large, broad at the base, and narrow at the tip—standing strongly upright. Each plant has a single stout flowering stem that supports a mass of large, deep blue flowers with petal lobes radiating away from puckered looking white centers. Like Jessica's stickseed, the seeds of this one cling to anything they touch—animal fur, socks, and pants. I photographed these along Rainier Road above Mule Creek and Highway 3 in mid-June.

WESTERN HOUND'S TONGUE
Cynoglossum occidentale
Borage family

5 symmetrical petals forming a tube

This hound's tongue is not as common nor as showy as its blue cousin but pretty none-the-less. This is a tall, hairy, multistemmed plant bearing loose, terminal clusters of brownish-purple flowers with yellow centers. Leaves taper to narrow bases. The fruits are distinctive nutlets about ½″ in diameter covered with little hooked barbs that stick to anything that rubs up against them—like socks and fur. I photographed this bunch on Montgomery Ridge above Trinity Lake in June.

PURPLE MAT, MATTED YERBA SANTA
Eriodictyon lobbii
Nama family

5 symmetrical petals forming a tube

I have run into purple mat in two location in the area covered by this guide. Once in July along the trail to Big Boulder Lake in the Trinity Alps and the other along Forest Highway 24 on the Trinity Divide. The population along Forest Hwy 24 was huge, forming extensive mats along a 100 yard section of the road shoulder. In both instances it was blooming on sandy, exposed, decomposed granitic soils. The plants are rangy with oblong leaves the length of the flowering stalk. The lilac to deep purple, tubular flowers bloom along the upper third of the stalk which often lies close to the ground. The stalks, leaves, and sepals are hairy.

PURPLE MILKWEED
Asclepias cordifolia
Dogbane family

GRAND HOUND'S TONGUE
Adelinia grandis
Borage family

WESTERN HOUND'S TONGUE
Cynoglossum occidentale
Borage family

PURPLE MAT, MATTED YERBA SANTA
Eriodictyon lobbii
Nama family

BLUEHEAD GILIA, BLUE FIELD GILIA
Gilia capitata
Phlox family

5 symmetrical petals forming a tube

This delicate blue gilia grows in troops in open, dry, rocky areas throughout the foothills in this area. The flowers bloom in compact clusters about 1″ to 1½″ across at the tops of 1′ to 2′ stems. Leaves are ½″ to 1½″ long, deeply lobed and mostly basal, with a few scattered along the lower portion of the stem. I shot this photograph on Musser Hill along the Weaver Basin Trail system in May.

ALPINE BREECHES, ALPINE WATERLEAF, WOOLEN BREECHES
Hydrophyllum alpestre
Waterleaf family

5 symmetrical petals forming a tube

A good place to find this little waterleaf is along the trail between Lion and Foster Lakes in the Trinity Alps and in the upper reaches of the Parker Divide. It closely resembles California waterleaf except for its size and location—growing at mid- to high-elevations in open, sunny sites—often blooming just as the snow is receding. The pale lilac flowers bloom in compact heads low to the ground nestled into plants that are usually no more than 8″ tall. The flowers are tinged with purple and have prominent deep lilac colored stamens and anthers. The leaves are bright green, spreading, deeply cleft and about 5″ to 6″ long. I photographed these just above Lion Lake in early June.

CALIFORNIA WATERLEAF, WESTERN WATERLEAF
Hydrophyllum occidentale
Waterleaf family

5 symmetrical petals forming a tube

California waterleaf much resembles its alpine cousin except for size and location. This one is much larger with plants often reaching 12″ or more in height and, unlike the alpine version, preferring shaded, damp sites at lower elevations. Flowers bloom in fairly compact heads are mostly white with ovetones of light purple. Stamens are prominent, darker purple topped and with gold anthers. The deep green leaves are up to 10″ long and spotted with light green giving the appearance of water spots, hence the name. I photographed these below the Hirz Mountain Lookout above Lake Shasta. I've also found it along Big French Creek and the lower end of the East Fork of Stewart's Fork in the Trinity Alps. An especially nice display is along Trinity Mountain Road on the Trinity Divide above French Gulch.

THREAD LINANTHUS
Leptosiphon filipes
Phlox family

5 symmetrical petals forming a tube

In keeping with my mantra of "you can never visit a place often enough if you're looking for wildflowers," I offer up this example. Over the past ten years I have walked through the Lowden Ranch Site—part of the Trinity River Restoration Project —hundreds of times and almost always always find something new to photograph. On this occasion I was photographing a beautiful Sierra morning glory when I spotted a population of thread linanthus scattered along a river access road. I had never seen them before. The tiny five-petaled, bluish-white flowers seemed to float over the ground on hair-like multi-stemmed stalks. The delicate petals are light blue, streaked with white, that radiates from a buttery yellow center. Stamens are prominent and topped with orange anthers. These photographs were taken on a flat just above the Trinity River near Lewiston in mid-June.

BLUEHEAD GILIA, BLUE FIELD GILIA
Gilia capitata
Phlox family

ALPINE BREECHES, ALPINE WATERLEAF
Hydrophyllum alpestre
Waterleaf family

CALIFORNIA WATERLEAF, WESTERN WATERLEAF
Hydrophyllum occidentale
Borage family

THREAD LINANTHUS
Leptosiphon filipes
Phlox family

SCOTT VALLEY PHACELIA
Phacelia greenei
Borage family

5 symmetrical petals forming a tube

Unless you're very attentive, you could walk over these tiny phacelias all day long and never know it. It's a serpentine loving species and especially abundant on Scott Mountain—one of my favorite hot spots for wildflowers. The tiny blooms are only 1/4" in diameter which open in coiled bunches atop stems that reach 2" to 3" in height. The flowers are deep purple with a stark white center. Stamens are fine, white, and topped with gold anthers. Photographing them is difficult when trying to get the camera lens low enough to capture an image. I've had to dig holes in the ground to lower the camera enough to get good images without disturbing the plants. I photographed these in late July on Scott Mountain.

SISKIYOU PHACELIA
Phacelia leonis
PRINGLE'S PHACELIA
Phacelia pringlei
Waterleaf family

5 symmetrical petals forming a tube

Sad to say, but even as attentive as I am to such things, I probably walked over dozens of these teeny wildflowers before I even realized they were there. This is one of the smallest and least obtrusive wildflowers I've ever photographed. They are so delicate that even breathing on them while making an exposures upset the entire shot. The leaves are narrower and more tapered than on the other phacelias depicted, mostly basal, and slightly hairy. The flowers are only 1/8" to 3/16" in diameter and a light blue or soft mauve (depending on the light). Stamens are long, thin, and white topped with gold anthers. Siskiyou phacelia is listed as rare and is endemic to the serpentine soils in the Klamath Ranges. I photographed the Siskiyou phacelia on the road to Cement Bluff Lake on the Trinity Divide. I photographed the Pringle's phacelia on Scott Mountain where I discovered they can put on quite a display—often blooming in large populations with Falcate onion. Pringle's phacelia is very similar, with slightly larger flowers, and is probably indistinguishable by most amateurs.

SCOTT VALLEY PHACELIA
Phacelia greenei
Borage family

SISKIYOU PHACELIA
Phacelia leonis
Borage family

PRINGLE'S PHACELIA
Phacelia pringlei
Borage family

CALIFORNIA POLEMONIUM, CALIFORNIA JACOB'S LADDER
Polemonium californicum
5 symmetrical petals forming a tube
Phlox family

This is a shade loving wildflower that often grows in large colonies beneath the canopies of fir and pine trees at higher elevations. It is a low growing plant with stems covered in soft hairs. The leaves are hairy, lance-shaped to somewhat oval, and pinnately divided into multiple leaflets. The delicate, light blue to light lilac, bell-shaped flowers bloom in loose clusters on stalks about 4" long, have prominent white stamens and a lemon-yellow throat. Before the flowers open the backsides of the petals are a beautiful periwinkle blue. The flowers are tube-forming but I like to think they look like a bowl when fully opened. Such are the vagaries of botany. I've photographed it in Onion Meadows, around Steveale Meadows, above upper Caribou Lake in the Trinity Alps, and above Upper Albert Lake in the Russians. In general, it is widespread throughout the Klamath Mountains.

WESTERN POLEMONIUM
Polemonium occidentale
5 symmetrical petals forming a tube
Phlox family

I found a fairly large population of this pretty polemonium blooming on a shaded, rocky slope as I was climbing to Upper Albert Lake in the Russians on an early September pack trip. The brilliant blue flower clusters caught my eye immediately, massed the way they were against the granite rock. The interior of the petals fade from soft blue at the outside edges to a creamy white and then to a golden-yellow and green at the throat. The outside of the petals fade from deep to light blue. The stamens and style are prominent with fat little white anthers. The deep green leaves are long and pinnately divided with 15 or more slightly hairy leaflets. Plants are, on average, about 12" to 14" tall.

SHORT JACOB'S LADDER, SKUNK-LEAVED POLEMONIUM
Polemonium pulcherrimum var. *pulcherrimum*
5 symmetrical petals forming a tube
Phlox family

This a small polemonium that loves open, dry, rocky, or sandy areas at higher elevations. The pinnately divided leaves are strongly linear, tough, and compact with small, egg-shaped leaflets. Some folks think they smell skunkish but, while they do have a strong odor, I wouldn't go as far as to compare the smell to that of a skunk —more like a chipmunk in bad need of a bath. The flowering stems are strongly erect with several small (½" across) periwinkle blue to light lilac flowers in a loose cluster at the top. The color in the throat varies from yellow to deep orange. The white stamens and style are prominent. I photographed these on the ridge above Little Duck Lake in the Russian Wilderness. I've also seen them above Upper Albert Lake in the Russians.

CALIFORNIA POLEMONIUM
Polemonium californicum
Phlox family

WESTERN POLEMONIUM
Polemonium occidentale
Phlox family

SHORT JACOB'S LADDER
Polemonium pulcherrimum
var. *pulcherrimum*
Phlox family

CALIFORNIA HAREBELL
Asyneuma prenanthoides
Bellflower family

5 petals joined into a bell

These diminutive little harebells are common at lower elevations and bloom in scattered bunches along tall, stiff stems that reach to 24". Each delicate blue flower has five narrow petals with deep purple mid-veins that arch away from its center—where you will find a long and conspicuous lilac colored style. They are partial to shade and common throughout these mountains. I photographed these on the Bailey Cove Trail above Shasta Lake. I've also found them at low elevations along the Stewart's Fork and Long Canyon Trails in the Trinity Alps, along the Paynes Lake and Duck Lake trails in the Russians, and around Whiskeytown lake.

ROUGH HAREBELL
Campanula scabrella
Bellflower family

5 petals joined into a bell

If this common name was up to me, I'd call it "tough" harebell just because it thrives in relatively dry and exposed habitats at higher elevations. I photographed these on the upper slopes of Mt. Eddy on the Trinity Divide where there is literally no shade and mid-afternoon temperatures get warm (see small photograph). Rocks surrounding the plants were hot to the touch but in spite of everything there they were—thousands of them with delicate blooms from light blue with hints of lilac to deep purple. Leaves are pubescent and stiff to the touch. Flowering stalks are likewise pubescent and rise to a height of 4" to 6" with a single upright bloom. Locally they are found along the north end of the Trinity Divide with a sighting in the limestones of Thurston Peaks.

WILKINS' HAREBELL
Campanula wilkinsiana
Bellflower family

5 petals joined into a bell

This little harebell is one of those rare wildflowers that, if attentive, you might find blooming in wet areas on serpentine soils in the Trinities and Marble Mountains. It is also found in the wet volcanic soils on Mt. Shasta. Interestingly enough, these are the only places on the entire planet where it exists. I've found it blooming next to melting snowbanks, the edges of streams, and small snow-melt ponds, springs, and seeps at higher elevations. The short flowering stems are strongly erect—each topped with a delicate lilac to purple flower with petals that curve outward with a protruding, purple style at the center. Several years ago I found a large and persistent population of a snow white phase in a meadow above Lake Anna. I photographed these in Long Canyon in the Trinity Alps in mid-July

CALIFORNIA HAREBELL
Asyneuma prenanthoides
Bellflower family

ROUGH HAREBELL
Campanula scabrella
Bellflower family

WILKINS' HAREBELL
Campanula wilkinsiana
Bellflower family

BLUEBELLS-OF-SCOTLAND
Campanula rotundifolia
Bellflower family

5 petals joined into a bell

Clusters of bright blue to light lilac, bell shaped flowers bloom in bunches along erect, hairless, 4″ to 20″ tall stems. The basal leaves are about 3″ long and narrow with smooth margins. The flower petals are pointed and flare slightly at the tips. Look for it in moist shaded areas at the edges of woods, on slopes and in meadows. This little Scottish immigrant is found throughout the northern latitudes in Europe. The Klamath Mountains are at the southern edge of its range in North America. I photographed this one along the Fen Trail above Kangaroo Lake on the Trinity Divide in June.

COMMON BLUECUP, VENUS' LOOKING GLASS
Githopsis specularioides
Bellflower family

5 petals joined into a bell

I have been taking pictures of the wildflowers on a particular slope above Shasta Lake for years and never noticed this beautiful bright blue species until spring 2020. Small and inconspicuous but when I put a macro lens on it and enlarged the picture—just like that I had another addition to this guide! The pretty cups stand straight up with slightly reflexed petals that are brilliant blue at the tips fading to white at the centers. Most of the flowers were a bit less than ½″ in diameter with several of them blooming on thin stalks rising from a common rootstock. Stems and leaves are hairy. Overall, the plant presents a gangly look. Although I photographed this in the foothills around Shasta Lake, there are populations along the North Fork of the Trinity River, Canyon Creek in the Trinity Alps, near Castle Crags State Park, and Whiskeytown National Recreation Area.

The Scott Mountains are an east/west trending range connecting the Red or eastern Trinity Alps with the northern terminus of the Trinity Divide—the "Transverse Range" of the north if you will. As you can see in the photograph, with the red and ochre rocky outcops and soils, this is serpentine country. It is home to myriad serpentine loving wildflowers including the bluebells of Scotland pictured above. Fens here are loaded with cobra lilies as well. This photograph was taken above Kangaroo Lake near the junction of the Fen Interpretive Trail and the Pacific Crest Trail which runs the length of the Scotts. I believe it was a huge failure back in the 1980s when this area was not considered for wilderness status—it is truly deserving of protection.

BLUEBELLS-OF-SCOTLAND
Campanula rotundifolia
Bellflower family

COMMON BLUECUP
VENUS' LOOKING GLASS
Githopsis specularioides
Bellflower family

The Scott Mountain Divide looking east towards China Peak

SISKIYOU DAISY, SISKIYOU FLEABANE
Erigeron cervinus
Sunflower family

flowers in composite heads

If you spend any time along Canyon Creek between the Sinks and the Canyon Creek Cascades you will find this beautiful little daisy in bloom—especially in July and early August. It doesn't like its feet in the water but definitely likes close proximity to it—finding a home in clefts in the granite outcrops just above the creek close to the spray zone. Flowers range in color from blue to pink, are about one to one and a half inches in diameter, and sit solitary on tall naked stalks. Leaves are dark green, somewhat leathery, narrowly oval, and lightly folded at the mid-rib. I photographed these above Middle Falls on Canyon Creek in mid-July.

DIFFUSE DAISY, SPREADING FLEABANE
Erigeron divergens
Sunflower family

Flowers in composite heads

So many little daisys—and so hard to ID! Alice Jones (*Flowers and Trees of the Trinity Alps)* and Bill Ferlatte (*A Flora of the Trinity Alps*) both made positive IDs and recorded locations in Big Flat and the Meadows around the Packer's Peak trailhead—which is where I found these. The flowers range in size from ¾" to 1" in diameter, are flat, with many narrow, pale white, blue, or lavender ray flowers. The disc flowers are deep yellow to almost orange. The plants are about 12" tall, somewhat rangy, and slightly spreading. Narrow, grayish-green leaves about ½" to 1" in length are arrayed along the stems. I found and photographed these growing in abundance along Coffee Creek Road in a dry meadow in full sun at the Packer's Peak Trailhead in early June.

MT. LASSEN FLEABANE
Erigeron lassenianus
Sunflower family

Flowers in composite heads

This tiny and colorful daisy blooms in profusion on the serpentine soil flats in the lower elevations of the Swift Creek and North Fork of Swift Creek areas. On one particular flat, where I photographed these, the population covered an area the size of a small house with delicate deep purple and blue fading to white flowers. The flowers range in size from smaller than a fingernail to larger than a thumbnail. They tend to grow in low mounds with the flowers opening on long stalks that lie flat to the ground. I photographed these alongside the road to the Lake Eleanor Trailhead in the Trinity Alps in May.

SISKIYOU DAISY
Erigeron cervinus
Sunflower family

DIFFUSE DAISY
SPREADING FLEABANE
Erigeron divergens
Sunflower family

MT. LASSEN FLEABANE
Erigeron lassenianus
Sunflower family

ANDERSON'S TUNDRA ASTER, ANDERSON'S MOUNTAIN-CROWN
Oreostemma alpigenum var. *andersonii* — Flowers in composite heads
Sunflower family

In my book (literally and figuratively) this aster is the most beautiful AND probably the most variable of all the asters found these mountains. Trying to identify it can drive an amateur nuts and that's why I've included the three color variations I've encounterd here on a page dedicated solely to them. Ranging from a beautiful powdery blue to a soft reddish-purple to snow white—they are all one in the same. They are found across a wide variety of habitats includng meadows, rocky slopes and flats, and the banks of snowmelt streams. They prefer being out in the full sun. They range in height up to two feet tall. Look for grass-like leaves about 6″ to 12″ long and 1/2″ wide. The flower heads occur individually at the tops of stems that may be a bit on the hairy side near the tops. The ray flowers can be anywhere from 1/2″ to 3/4″ long. The disc flowers are always yellow to orange. I photographed the blue form below the outlet of Smith Lake in late August, the purple variation in the meadows surrounding L Lake in late September, and the white variation in the meadows behind Upper Caribou Lake in late August—all in the Trinity Alps. The white variation is rare. In addition to the Caribou population the only other place I've encountered it was on the slopes above Salmon Lake, also in the Trinity Alps.

ANDERSON'S TUNDRA ASTER, ANDERSON'S MOUNTAIN-CROWN
Oreostemma alpigenum var. *andersonii*
Sunflower family

CASCADE ASTER
Eucephalus ledophyllus var. *covillei* Flowers in composite heads
Sunflower family

This aster is easily recognized by the narrow petals and blue to lavender ray flowers. The petals can be droopy, or recurved, and surround the small, compact head of orange disc flowers. Stamens and anthers are prominent. The flowering stems, each with a single flower head at the top of each branch, can be up to 2′ tall. It tends to grow in scattered colonies but I have found it in thick, compact groups of 20 or more. They are common to these mountains throughout the Trinities as well as the Russians. They prefer well watered sites and partial shade. I photographed these on the rocky slopes above Lower Russian Lake in the Russian Wilderness in September.

TALL STEPHANOMERIA, WAND WIRELETTUCE
Stephanomeria virgata ssp. *pleurocarpa* Flowers in composite heads
Sunflower family

The individual flowering stalks are tall—some approaching two feet —with seed pods dry fruits and flower heads crowded along the upper half. Mature blooms have five ray flowers that are strap shaped and narrow at the ovary flaring outward towards the middle before slightly narrowing again at the tips. The color is a beautiful soft mauve. When I found this pretty wildflower there were only half a dozen plants in the vicinity and most of the flowers had gone to seed. I will be re-visiting this one when they are in their prime during the next field season. I shot these photographs at the spring above the Maple Mine along the trail to East Fork Lakes in the Trinity Alps in mid-August.

WESTERN MOUNTAIN ASTER
Symphyotrichum spathulatum Flowers in composite heads
Sunflower family

This is an aster common in damp meadows, the shorelines of lakes, ponds, and streamsides throughout the high country. Flower heads bloom in loose clusters at the tops of erect and hairy stems. Basal leaves are 2″ - 6″ long and narrow. Leaves growing up the stems are much smaller. The narrow ray flowers range in color from light lavender to deep purple and are surround by a dense cluster of yellow to brownish disc flowers. I photographed these in a wet meadow along the Pacific Crest Trail between Doe and Little Granite Lakes in August. I have also found them blooming in little pocket meadows on the granite slicks between Smith Lake and Morris Meadows all in the Trinity Alps.

CASCADE ASTER
Eucephalus ledophyllus var. *covillei*
Sunflower family

TALL STEPHANOMERIA
WAND WIRELETTUCE
Stephanomeria virgata
ssp. *pleurocarpa*
Sunflower family

WESTERN MOUNTAIN ASTER
Symphyotrichum spathulatum
Sunflower family

DEER BRUSH, WILD LILAC
Ceanothus integerrimus
Buckthorn family

5 symmetrical petals - ovary inferior

When most people think of ceanothus (if they think of ceanothus at all), this is the one that comes to mind. Unlike the evergreen wedgeleaf and Lemmon's ceanothus, this is a deciduous shrub. It grows in huge thickets in open areas following fire or other disturbance throughout the lower elevations in this area, and puts on magnificent displays of white, pinkish-lilac, or light blue flowers in pyramidal, lilac-like clusters. It is a stiffly branched shrub that often grows to 10′ in height. It is a much favored over-wintering food of deer and host for deer ticks. If you want to see a beautiful display of wild lilac just drive Highway 299 over Oregon Mountain west of Weaverville. Following the wildfires between Junction City and Weaverville in 2016 it has literally taken over the entire mountainside. I photographed these along Van Ness Road above Trinity Lake in May.

LEMMON'S CEANOTHUS
Ceanothus lemmonii
Buckthorn family

5 symmetrical petals - ovary inferior

This ceanothus puts on a beautiful display of vibrant, deep blue flowers early in the season from the foothills of the Sacramento Valley to mid-elevation trails in the back-country. It is a stiffly branched shrub that sometimes reaches 3′ in height, with light gray-white bark. The tiny flowers bloom in tight rounded clusters about 2″ or so in diameter. Leaves are small, alternately arranged, evergreen, oblong to almost round, tapered at the base, and finely gland-toothed. I photographed this one along the road to the North Fork of Swift Creek Trailhead in late May.

YERBA SANTA, KNIT-BONE
Eriodictyon californicum
Nama family

5 symmetrical petals forming a tube

This is a leggy shrub that likes hot, dry, open areas. It is often one of the first to colonize an area after low elevation wildland fire. The leaves are narrowly lance shaped, tough, leathery, toothed and, in the dead of summer, coated with a sticky, highly flammable resin that glazes them and creates a shiny appearance. It forms dense thickets with individual shrubs reaching 6′ or more in height. The pink tinged white tubular-shaped flowers are pretty and bloom in an irregular head at the tops of the branches. By late summer, depending on location, a majority of the shrubs have turned an unattractive brownish green with tall spikes of brown seed-heads. The common name "knit-bone" comes from old medicinal lore indicating that drinking tea made from the leaves would help broken bones heal. After suffering a shattered heel years ago I was presented with a rather potent drink at regular intervals until my cast was removed and I could hobble around again. I have no clue about its actual effectiveness but I did as I was told and everything healed just fine. There have been no lasting consequences, except having absolutely no desire to ever drink the stuff again! When Julie Knorr (one of the editors of this guide) was in college, she and her friends would chew on the leaves ("terrible" she says) and then drink water, which then tasted like it had been sweetened. "I don't know why we did that, or anything else we did in college in the 70's" She said!

DEER BRUSH, WILD LILAC
Ceanothus integerrimus
Buckthorn family

LEMMON'S CEANOTHUS
Ceanothus lemmonii
Buckthorn family

YERBA SANTA, KNIT-BONE
Eriodictyon californicum
Nama family

MAHALA MAT

Ceanothus prostratus

Buckthorn family

5 symmetrical petals - ovary inferior

Mahala mat is common throughout these mountains in dry, semi-exposed sites where it forms dense, low growing mats 12″ to 14″ in height. The flowers are grouped in tight, sweet smelling clusters and are variable in color. They range from a light violet (almost white) to deep blue with several shades in between. The leaves are small, stiff, and spiny. It's an attractive plant and I've known people who transplanted starts hoping for ground cover in their yards, but most attempts have been unsuccessful. A better bet would be to buy starts at a native plant nursery. I photographed the light violet version (top) alongside the road to the Granite Peak trailhead, the deep violet version (center) along the Scorpion Creek Road to Bonanza King, and the dark blue version (bottom) along the road to the Lake Eleanor Trailhead all in May. The side-bar photograph shows Mahala mat carpeting the edges of the Stoney Ridge trail in the Trinity Alps.

MAHALA MAT
Ceanothus prostratus
Buckthorn family

Three common color variations

Monument plants (*Frasera speciosa*) decorate Back Meadows in the Marble Mountain Wilderness.

Green corn lily, arrowleaf ragwort and western yarrow fill a niche in a granite outcrop above Middle Caribou Lake in the Trinity Alps

BROWNIE LADY'S SLIPPER, CLUSTERED LADY'S SLIPPER
Cypripedium fasciculatum 3 or 6 petals
Orchid family

This is a small and inconspicuous orchid that blends nicely into the background of the forest floor. The two large, parallel veined leaves are the most prominent feature and what I generally notice well before I see the flowers—which are often hidden beneath them. The little reddish-green flowers bloom in clusters on a stalk rising from between the leaves. The plant ranges to about 5" tall. **Note:** In a paper recently published by the Institute for Applied Ecology, Dr. Thomas Kaye in conjunction with the Bureau of Land Management, noted that due in part to climate change, roughly half the populations of this orchid are likely extinct. Lower elevation populations seem to be most at risk. This is a definite concern and reinforces the fact that they should be left alone and not disturbed—true for all orchids.

STREAM ORCHID, GIANT HELLEBORINE, CHATTERBOX
Epipactis gigantea 3 or 6 petals
Orchid family

As the name implies, this orchid favors mid-elevation springs, seeps, fens, and streamsides. They are found in large colonies and can range from a few inches to 3' in height. Flowers are borne in the leaf axils with three long, reddish-green sepals. The upper two petals are yellowish-green with prominent blood-red veins. The lower petal droops and extends outward with a deep yellow to white tongue, lightly veined with red. They are still easily passed by because they blend in with the surroundings. I photographed these along the Parks Creek Road on the west side of the Trinity Divide. **Note:** I am always intrigued with the common names assigned to the different wildflowers and "chatterbox" is no exception. After doing a bit of research I found that the lower "lip" flutters up and down when flies pollinate the flower, hence the name chatterbox!

CHECKER LILY, MISSION BELLS
Fritillaria affinis 3 or 6 petals
Lily family

I have photographed this pretty fritillary in several locations in the limestone around the McCloud Arm of Shasta Lake. The best display I've found was on Hirz Mountain where it was abundant along the road to the lookout. You really have to look for it though, as it disguises itself well in the surrounding vegetation. The flowers are a dark purple-brown with petals long and angular. The anthers are a deep brick red. Several nodding flowers bloom up 12" - 14" flowering stalk. The long, oval leaves form a whorl flat to the ground with another whorl further up the stalk. These photographs were taken in April.

MOUNTAIN BELLS, SPOTTED MOUNTAIN BELLS, MOUNTAIN FRITILLARY
Fritillaria atropurpurea 3 or 6 petals
Lily family

These fritillaries are found in scattered populations on drier soils. The mottled yellow and reddish-brown petals blend nicely with the surroundings and even a seasoned wildflower hunter can walk past them! The flowers are nodding and bell shaped with prominent golden-yellow anthers. Flowering stalks emerge from groups of mostly upright, long, narrow leaves. Smaller leaves are scattered along the flowering stalk. Each stalk can sport several flowers. The petal surface appears dull from above; but lie down in the dirt and look up through the petals from beneath, back-lit by the sun, and they are a brilliant yellow and red! I photographed these in the meadows surrounding East Boulder Lake in early June.

BROWNIE LADY'S SLIPPER
Cypripedium fasciculatum
Orchid family

STREAM ORCHID, CHATTERBOX
Epipactis gigantea
Orchid family

CHECKER LILY, MISSION BELLS
Fritillaria affinis
Lily family

SPOTTED MOUNTAIN BELLS
Fritillaria atropurpurea
Lily family

GREEN CORN LILY
Veratrum viride 3 or 6 petals
False-hellebore family

Green corn lilies are more commonly found from the northern Trinity Alps, Russian, and Marble Mountain wilderness areas where they are widely interspersed with their white-flowered cousin the California corn lily. True to their name the six-petaled flowers are green and bloom on a loose array of drooping stems. Bright yellow and sometimes blue spatulate shaped anthers (individual flowers can share both) decorate the centers of each flower. Individual plants reach 5' - 6' in height and are found in large colonies. I photographed these below Upper Albert Lake in the Russian Wilderness. I've also photographed them below South Fork Lakes, Telephone Lake, and Caribou Lake—all in the Trinity Alps.

MONUMENT PLANT, DEER EARS
Frasera speciosa - Swertia radiata 4 petals
Gentian family

This is the giant in the gentian family. With some plants ranging to 5' tall, it is impossible to miss—even by the most distracted backpacker or day hiker. Thick flowering stalks rise out of masses of large whorled elliptic leaves that are pointed at the tips. More leaves occur in whorls along the stalk growing smaller as they progress upwards. Masses of beautiful flowers bloom between the leaf whorls and open progressively from the bottom to the top. Petals are yellowish-green, fading to white at the center, dotted with deep purple with sharply folded and pointed tips. Petal margins are silvery or light purple. Each petal has two purple and hairy looking nectary pits with long, white deeply divided scales extending outward from the base of the ovary to cover them. Stamens are prominent, stout, upright, and tapered with large yellow to orange anthers. The dozens of plants I found were popular with ants foraging in the nectary pits. Monument plants are monocarpic, meaning that they grow for several years before they flower. After flowering and setting seeds they die. In any area where they grow you can find a mix of dead stalks, flowering plants, and others in various stages of growth. They were quite common in Big and Back Meadows as well as the meadows surrounding the Wright Lakes in the Marble Mountains. I photographed these in late July.

Wright Lakes in the Marble Mountain Wilderness: These lakes are literally surrounded by wildflower gardens and when I visited here in late July the bloom was in its absolute prime. They are lush and inviting. Spend some time roaming around here and you will want to return often. I guarantee it.

GREEN CORN LILY
Veratrum viride
False-hellebore family

MONUMENT PLANT, DEER EARS
Frasera speciosa - Swertia radiata
Gentian family

The meadows above Lower Wright Lake in the Marble Mountains

CALIFORNIA PIPEVINE
Aristolochia californica 3 or 6 petals
Pipevine family

Here's one that looks more like a little green duck out of water than a wildflower. It is and closely related to the Dutchman's pipe of the eastern U.S. It is among the earliest of bloomers in the lower elevations, often appearing in late January or early February in the Sacramento Valley and mid-March in Trinity County. The green and coppery-red flowers open before the leaves and hang in thick masses on twisted, tough, vining stems that climb into surrounding brush, and small trees. The heart-shaped leaves and stems are fuzzy and bright green. I photographed these in the limestones above Shasta Lake but look for them in more shaded sites at lower elevations along the Trinity River and other drainages.

CALIFORNIA GREENBRIAR, GREENBRIAR
Smilax californica 3 or 6 petals
Smilax family

This greenbrier is similar in almost all ways to its close cousin English Peak greenbrier with two obvious exceptions—the first being its preferred habitat. This one thrives in shade but does not seem to need a particular close association with water. The second and most important difference is that the stems of this one are enveloped in stiff thornsprickles. It is a pretty vine that can be found twining its way through brush and into small trees. Leaves are heart shaped and prominently parallel-veined. Flowers are tiny, delicate and fall apart at the slightest disturbance which can pose a problem when trying to photograph them. The sepals and petals are indistinguishable from each other and are actually called "tepals." They bloom in umbels along the entire length of the vine. Dark, almost black pea-sized fruits develop later in the season. I photographed this one on a shaded flat along the Upper Trinity River in mid-June.

ENGLISH PEAK GREENBRIER
Smilax jamesii 3 or 6 petals
Smilax family

This climbing vine likes lots of water and partial to deep shade. You will find it in alder thickets along streambanks, or near springs, seeps or other damp places. I've found it in abundance along the Stewart's Fork, Bowerman Meadows, and Poison Canyon Trails in the Trinity Alps. It climbs on herbaceous stems to heights of 8' or 9'. Leaves are large and heart-shaped with three to five main veins converging at the leaf base. The tiny and delicate flowers have petals and sepals that are indistinguishable from each other, called 'tepals', which bloom in umbels along the length of the vine. The flowers are so delicate that I found it difficult to photograph them. Just breathing on the flowers or the slightest movement made them scatter like crazy and fall to the ground. The vines are smooth and spineless. Bunches of pea sized fruit develops in late spring and ripens in late summer to a deep black. I photographed the flowers along the Poison Canyon Trail in mid-June and the fruit along Stewart's Fork in late August.

CALIFORNIA PIPEVINE
Aristolochia californica
Pipevine family

CALIFORNIA GREENBRIAR
Smilax californica
Smilax family

ENGLISH PEAK GREENBRIAR
Smilax jamesii
Smilax family

SPARSE-FLOWERED BOG ORCHID
Platanthera spoarsiflora
Orchid family

3 or 6 petals

This little green orchid often hides in plain sight—overshadowed by other plants in wet meadows, near seeps, springs, and stream banks. This is one of those delicate little "belly flowers" that rarely exceed 12" in height a. My Granddaughter claims the tiny blooms look like fish with their mouths open and observant hikers will likely agree. The left hand photograph was taken in a wet seep below Little Duck Lake in the Russians and the one at right alongside a stream in Doe Flat in the Trinity Alps.

LONG-TAILED GINGER
Asarum caudatum
Pipevine family

3 or 6 petals

Distinguished by the long tails on each sepal (there are no petals), this deep reddish brown flower hides beneath a low canopy of its own bright green, heart shaped leaves which do not exhibit the pale veins of Hartweg's seen below. I photographed this one in Poison Canyon, alongside the trail to Lilypad Lake in late June. I have also found it along the Stewart's Fork Trail in forested areas above Morris Meadows. Alice Jones reports it from Boulder Creek in the Trinity Alps. It gets more common the farther west you go in Trinity County. It likes damp and deeply shaded sites.

HARTWEG'S WILD-GINGER
Asarum hartwegii
Pipevine family

3 or 6 petals

If you look closely at the sepals (there are no petals) you will see they can be a deep brick red or pale reddish-green. The tails on the sepals are pronounced but not nearly as long as those of long tailed ginger. Leaves are heart shaped, deep green with pronounced white venation. When I was a kid I woulds help my Grandmother harvest the roots which she would clean, cut into bite-sized chunks, and boil lightly in a sugar solution. While still slightly damp she would roll the pieces in coarse sugar and place them in a candy dish on her table. They were a bit tough but delicious and spicy. Our wild ginger is not botanically related to the fresh or powdered ginger we buy at the grocery store, though the scent and flavor are

SPARSE-FLOWERED BOG ORCHID
Platanthera sparsiflora
Orchid family

LONG-TAILED WILD GINGER
Asarum caudatum
Pipevine family

HARTWEG'S WILD GINGER
Asarum hartwegii
Pipevine family

The woods and manzanita patches along the trail to little Boulder Lake in the Trinity Alps is home to good populations of Pinedrops and California groundcones.

Striped coralroot - *Coralloriza striata*

PHANTOM ORCHID
Cephalanthera austiniae
Orchid family

3 or 6 petals

Phantom orchids are not only gorgeous but they are also predictable—growing in the same place year after year. The creamy white plants grow to 10" in the deep shade of mixed conifer forests at lower elevations in the mountains. The stems, sepals, and petals are all white except for a small patch of deep yellow orange on the lower lip. Some people confuse these with ghost plants but the two are easily distinguished. Flowers of the ghost plant lack the yellow patch on the lower petal, hang downward like bells and the petals are arranged regularly around the superior ovary, while those of the phantom orchid face outward and upward with irregular flower parts attached at the tops of inferior ovaries. Because of their tendency for inhabiting deep shaded forest, it is fairly easy to pass them by. I photographed these above the Long Canyon Trail in late June.

FRINGED PINESAP
Pleuricospora fimbriolata
Heath family

5 symmetrical petals - ovary superior

Years ago I photographed this plant near Josephine Lake in the Russians and years later I ran into it again, this time on the trail to Duck Lakes. My grandkids actually found them blooming beneath a dense canopy of white fir and weeping spruce saplings in late August. The plants below Josephine Lake were typical for this species, about 6" tall with pinkish-white stems and red, scale-like clasping leaves. The flowers are distributed around the top of the stems and point outward and upward. The Duck Lakes examples were ground huggers, the stems buried in duff with all of the flowers arrayed evenly around the tops of the stems, pointing outward and upward. In both cases the flowers were creamy white, bell shaped, and flared at the tips.

GHOST PLANT, INDIAN PIPE
Monotropa uniflora
Heath family

5 symmetrical petals - ovary superior

I prefer ghost plant to Indian pipe because, in this mountainous region, they are harder to find than ghosts in the closet! I've seen it twice in 50 years—once near Slate Mountain on the Trinity Divide and once at Lipstick Lake in the Russian Wilderness. Alice Jones found it along Boulder Creek in the Trinity Alps. When fresh, these gregarious plants have bell-shaped, translucent white flowers tinged with pink. The stalks are likewise translucent white and 6" to 10" tall. Plants begin to develop black spots as they age until, eventually, the entire plant turns black and withers. They prefer deep duff in conifer forests. I photographed these in Redwoods National Park in early July. If you do happen to find one, take a photograph, record the location, and let a Forest Service botanist know. Your data would be greatly appreciated!

PHANTOM ORCHID
Cephalanthera austiniae
Orchid family

FRINGED PINESAP
Pleuricospora fimbriolata
Heath family

GHOST PLANT, INDIAN PIPE
Monotropa uniflora
Heath family

PINEDROPS
Pterospora andromedea
Heath family

5 symmetrical petals - ovary superior

This is a giant among mycotrophs, often reaching heights of five feet. It is common in the understory of mixed conifer forests at mid to upper elevations throughout this area and, because of its size, easily recognized. The stalks are stout, reddish brown, and leafless. Masses of deep orange-yellow bell-shaped flowers nod downward from the ends of short red stems on the upper fourth of the plant. The entire plant is covered with sticky, short hairs. Even after the flowers have faded, the stalks and dry seed pods are still evident, sometimes lasting into the next flowering season. I photographed these along the Swift Creek Trail in the Trinity Alps.

CANCER ROOT, CLUSTERED BROOMRAPE
Aphyllon fasciculatum
Broomrape family

5 irregular petals forming a tube

Following a large wildfire near Shasta Lake, cancer root emerged during the blooming season in huge clusters. It Is a mystery to me why someone would choose to name such a beautiful little flower as they did. Maybe it's because they suck the life out of their hosts! At any rate, I find them delightful and enjoy the deep reddish-brown, erect, hairy stems that support a single, tubular mustard-yellow flower. Even though they are small, they are hard to miss. I photographed these along the road to the Lake Eleanor and Shimmy Lake Trailhead in the Trinity Alps in mid-May.

NAKED BROOMRAPE
Aphyllon purpureum
Broomrape family

5 irregular petals forming a tube

Because of their size, these exquisite members of the broomrape family are often difficult to find. Those that I've been fortunate enough to photograph have been small—only 2″ to 3″ in height and often blending into their surroundings. The greenish-brown stems are erect and covered with sticky hairs. Each stem supports a single vibrant blue flower with deep orange-yellow in the throat. Lower parts of the petals are mustard yellow where they emerge from the calyx. I found it blooming along the shorline of Papoose Lake in late August and in large numbers in association with broad-leaf stonecrop and Oregon saxifrage along Canyon Creek Road in mid-April. I've found scattered individuals in no apparent association with other plants in open and exposed areas on Scott Mountain. I shot the large photograph along Canyon Creek and the inset on Scott Mountain in July.

PINEDROPS
Pterospora andromedea
Heath family

CANCER ROOT, CLUSTERED BROOMRAPE
Aphyllon fasciculatum
Broomrape family

NAKED BROOMRAPE
Aphyllon purpureum
Broomrape family

SPOTTED CORALROOT
Corallorhiza maculata
Orchid family

3 or 6 petals

Spotted coralroot is a common trailside mycotroph found throughout the area covered by this book. This one is distinguished by its overall deep reddish-brown coloration. The sepals are dark reddish-brown around the edges and yellowish brown near the center with dark striations running the length. The lateral petals are much the same as the sepals. The broad lip is snow white with deep maroon spots and ruffled tips. The prominent column, a deep buttery yellow with small maroon spots, consists of a fused style and stamens and curves upward and outward from the throat. I understand that lemon yellow plants, lacking the reddish pigment, are not uncommon. The plants tend to grow in large clusters in deep shade under conifers. I have photographed these on the Long Canyon Trail in the Trinity Alps and lower down along Swift Creek just below Highway 3 in early July.

STRIPED CORALROOT
Corallorhiza striata
Orchid family

3 or 6 petals

This is a beautiful coralroot that is sometimes hard to find. The species prefers shaded, damp sites with knee deep forest litter. The first time I encountered it was during a heavy snowstorm in late May as I was hiking from Emerald Lake back to Morris Meadows with niece and nephew in tow. Photographing it at the time was questionable so, several years later, when this guide began to take shape, I went on the hunt for it again. I returned to the same area and found several plants scattered in tangled masses of downed branches and other woody debris. The dark reddish brown stems grow to 10″ tall and host several candy-striped flowers along the upper half. The lower lip is deep maroon while the sepals and lateral petals are lighter with deep maroon stripes. The column is a vibrant buttery yellow. This photograph was taken above Morris Meadows on the trail to Emerald Lake in early June.

WESTERN CORALROOT, MERTENS' CORALROOT
Corallorhiza mertensiana
Orchid family

3 or 6 petals

Like some of the other coral roots, this one has a reddish pink stem and stands 10″ to 14″ tall. The orchid flowers are scattered alternately along the stem and face outward. The lateral petals are maroon and yellow with small maroon spots at the tips. The reddish sepals are narrow and elongated. The lower lip is much broader, whitish rimmed, with deep lavender. A prominent yellow column curves upwards and out over the throat. Lemon yellow variations are not uncommon but I profess to having found them only once. They were entirely yellow except for the lower lip which was snow white. I photographed them alongside the Stewart's Fork Trail in mid-August. I photographed the bunch on the left along the trail just below Doe Lake in September—which is fairly late for coral roots.

SPOTTED CORALROOT
Corallorhiza maculata
Orchid family

STRIPED CORALROOT
Corallorhiza striata
Orchid family

WESTERN CORALROOT, MERTENS' CORALROOT
Corallorhiza mertensiana
Orchid family

SUGARSTICK, CANDYSTICK
Allotropa virgata 5 symmetrical petals - ovary superior
Heath family

Here is a strikingly beautiful mycotroph found commonly throughout the area on granitic soils. The tall (up to 15") red and white vertically striped stalks are, at first, wrapped with white bracts that furl outward revealing small, bowl shaped white flowers with a mass of short dark stamens. Typically, they grow in clusters One of the nicest populations I've ever seen was at Eaton Lake in the Russians in early July. They are also quite common in the Sugar, Duck, and French Creek drainages. This photograph, however, was taken at Crater Lake in south-central Oregon.

SNOW PLANT
Sarcodes sanguinea 5 symmetrical petals - ovary superior
Heath family

This is definitely the flashiest of all the mycotrophs and, because of the size and brilliant red color, are hard to miss. The robust plants reach 12" in height and achieve deep red throughout because of the numerous red flowers. They bloom in large, extended colonies. At one time in California (back in the 1950s and 60s), I remember hearing that these plants were protected and that there was a $500 fine for picking them but I am uncertain if that was even true (I was 10 at the time). They are actually quite common at mid elevations. I photographed these along the trail to Stoddard Lake in early July. Look for them around Eaton Lake in the Russians where my grand-kids spotted them blooming on a decomposed granite flat way above the opposite shoreline more than 300 yards away.

GNOME PLANT
Hemitomes congestum 5 symmetrical petals -ovary superior
Heath family

It's hard to miss this one if it is within eye-shot of a trail because of the vibrant pink to white blooms. It first appears as a small, rounded mound of almost yellowish-salmon pink flowers on the forest floor in deep litter. As the flowers open they exhibit a translucent pink vibrancy. Individual plants are generally 3" - 6" in diameter. The name gnome plant is entirely fitting because of the size and the shaded locations where it is found. It's easy to identify because it does not look at all like any of the other mycotrophs.

The first time I ever laid eyes on these was next to a spring of ice cold water along the trail to Eaton Lake in the Russians (one of those springs where you get eaten alive by mosquitoes if you stop to take photographs!). For successive years after, I would look in the same place but I never found it there again. It was years later, while exploring for mushrooms and lichens in Patrick's Point State Park on the North coast that I came across another population. Later, my grandson and I stumbled onto a colony growing along the Boulder Creek Trail just above Boulder Creek in the Trinity Alps. That was in late August and is where I took this photograph.

SUGARSTICK, CANDYSTICK
Allotropa virgata
Heath family

SNOW PLANT
Sarcodes sanguinea
Heath family

GNOME PLANT
Hemitomes congestum
Heath family

CALIFORNIA GROUND-CONE
Kopsiopsis strobilacea
Broomrape family

5 irregular petals forming a tube

Spend time in a patch of manzanita or around a grove of madrones and chances are good you will find California ground-cones. At first glance, they appear to be pine cones lying about. Young ground-cones are somewhat fleshy and pliable and have gorgeous purple flowers. As they age, they dry out and turn brown and brittle. At lower elevations the 'cones' can be quite large (9" tall). At higher elevations, say along the trail to Duck Lake or around Eaton Lake in the Russians, where I photographed the brownish-purple one, they rarely exceed 5" or 6" in height. The yellow variation, which I photographed on the ridge between Josephine and Horseshoe Lakes in the Russians was, likewise, only about 6" tall. Look closely for them in May in the foothills or in July higher up.

A colony of California ground-cones found in association with madrone along Canyon Creek in the Trinity Alps. This is a prime example of a parasitic plant which was a small part of a large population spread out over an eighth of an acre.

CALIFORNIA GROUND-CONE
Kopsiopsis strobilacea
Broomrape family

DWARF MISTLETOE

Arceuthobium ssp. no obvious petals
Mistletoe family

There are approximately twelve taxa of dwarf mistletoe (*Arceuthobium* spp.) found commonly on conifers like true fir (*Abies* spp.), incense-cedar (*Calocedrus decurrens*), pine (*Pinus* spp.), Douglas-fir (*Pseudostuga menziesii*), and juniper (*Juniperus* spp.) in the Klamath Mountains. These are parasitic in nature, tapping into the vascular system of the tree and draining nutrients and water. Late in the season, if you lightly pinch the fruit between your fingers they explode with a popping noise that kids and inquisitive adults find fascinating. They are often called exploding booger plants because the seed coats are sticky and, once ejected, fly and stick to a new host tree. The photograph at left was taken near Lewiston and the photograph on the right below Bear Gap at the lower end of Morris Meadows in the Trinity Alps.

OAK MISTLETOE, AMERICAN MISTLETOE

Phoradendron leucarpum subsp. *tomentosum* No obvious petals
Mistletoe family

This is a hemi-parasitic plant commonly found growing in oak trees and occasionally manzanitas throughout the area. As a hemi-parasite, they derive some energy from photosynthesis and the rest they take from their host. Phorodendron is characterized by large, stiff, gray green to olive-green branches with numerously paired, oval, stiff, coarse leaves. Mistletoe probes the host tree for nutrients and eventually leads to the decay of the branches it has attached itself to. The plants are dioecious, producing either male or female flowers in long clusters. Female flowers yield pink to salmon colored berries. A similar species, *Phoradendron juniperinum*, is found on conifers across the region. I photographed these with the berries where they had fallen beneath black oaks above Lewiston Lake in early May, The small photograph was taken in the foothills above Redding. **Note**: Dioecious means "two houses" in Greek. Male and female plants are separate and any one plant will produce only male or female flowers.

Mistletoe magic... Kissing under sprigs of mistletoe is a well-known holiday tradition, but this little plant's history as a symbolic herb dates back thousands of years. Many ancient cultures prized mistletoe for its healing properties. The Greeks were known to use it as a cure for everything from menstrual cramps to spleen disorders, and the Roman naturalist Pliny the Elder noted it could be used as a balm against epilepsy, ulcers, and poisons. The plant's romantic overtones most likely started with the Celtic Druids in the 1st century A.D. Because mistletoe could blossom even during the long, dark frozen winter, the Druids came to view it as a sacred symbol of virility, and they administered it to humans and animals alike in the hope of restoring fertility. The Anglo Saxon translation for mistetoe is "dung twig," because they noted that it grew where perching birds did their "business." Mistletoe's associations with fertility and virility continued through the Middle Ages, and by the 18th century it had become widely incorporated into Christmas celebrations. Just how it made the jump from sacred herb to holiday decoration remains up for debate, but the kissing tradition appears to have first caught on among servants in England before spreading to the middle classes. As part of the early custom, men were allowed to steal a kiss from any woman caught standing under the mistletoe, and refusing was viewed as bad luck. Yet another tradition instructed the merrymakers to pluck a single berry from the mistletoe with each kiss, and to stop smooching once they were all gone.

DWARF MISTLETOE
Arceuthobium spp.
Mistletoe family

OAK MISTLETOE
Phoradendron leucarpum subsp. *tomentosum*
Mistletoe family

Index of Common Names

Index of Scientific Names

Contributors

Ken DeCamp was born to parents who took him on his first backpacking trip when he was 4 months old into Glacier National Park. Since that time he has logged thousands of trail miles in the Pacific Northwest, Alaska, the Carolinas, Georgia, Pakistan, Australia, and Switzerland. He has been drawing and photographing wildflowers for over 40 years and it is from this collection of images that this book became a reality. Though he has travelled widely, he has always considered the mountains of northern California his home and has, for 60 years, explored their most hard to reach places. He retired from the USDA Forest Service in 2008 after a 38 year career in Fire, Land Management Planning, and Public Affairs. He lives with his wife Pam in Shasta Lake where they spend time with their family and their only pet, a cardboard dog named "Flat fido," gifted by a friend who thought that, someday, they might need serious protection from something. Ken is also an avid backpacker, trail runner, and mountain biker.

Julie Kierstead has been a working botanist since 1976, doing rare plant surveys and conservation work in Oregon and California. She has a B.S. in botany from Oregon State University and an M.S. in biology from Northern Arizona University. She worked for the Berry Botanic Garden in Portland as conservation director, developing a seed bank for rare & endangered plants of the Pacific Northwest and helped get a state endangered species act passed in Oregon. Since 1989 she has had the good fortune to serve as Forest Botanist for the Shasta-Trinity National Forest in northwestern California, where she is ideally situated to promote plant exploration in the Klamath Ranges and southern Cascades. This is not her first venture into the field of publishing having served as co-author and technical editor for the University of California Agriculture & Natural Resources publication ***Illustrated Field Guide to Selected Rare Plants of Northern California*** published in 2001.

Julie Knorr retired as a botanist after a 33 year career with the U.S.D.A. Forest Service on the Klamath National Forest. She grew up in Colorado where she gained a lifelong love of the mountains from her biologist parents. In 1978, after receiving a B.S. in botany from University of California at Davis, she started work on the Klamath N.F. During her career as a botanist, she worked primarily in the areas of rare plant protection, vascular plant and lichen monitoring, habitat restoration, and environmental interpretation. Post retirement, she has continued to do volunteer and contract surveys for the Klamath National Forest, primarily in the Marble Mountain, Russian, and Trinity Alps Wilderness areas. She lives along the Scott River in the home she shared with her late wife, Marsha, in the shadows of the Marble Mountain Wilderness in Siskiyou County, California. She now spends much of her time viewing the world from a bicycle.